VOLUME FIFTY ONE

THE ENZYMES

DNA Damage and Double Strand Breaks - Part A

Serial Editors

LAURIE S. KAGUNI

Department of Biochemistry and Molecular Biology, Michigan State University, East Lansing, Michigan, United States

FUYUHIKO TAMANOI

University of California, Los Angeles, USA

VOLUME FIFTY ONE

THE ENZYMES

DNA Damage and Double Strand Breaks - Part A

Edited by

FUYUHIKO TAMANOI

Institute for Integrated Cell-Material Sciences, Institute for Advanced Study, Kyoto University, Kyoto, Japan

KENICHI YOSHIKAWA

Faculty of Life and Medical Sciences, Doshisha University; Center for Integrative Medicine and Physics, Institute for Advanced Study, Kyoto University, Kyoto, Japan

ELSEVIER

ACADEMIC PRESS

An imprint of Elsevier

Academic Press is an imprint of Elsevier
50 Hampshire Street, 5th Floor, Cambridge, MA 02139, United States
525 B Street, Suite 1650, San Diego, CA 92101, United States
The Boulevard, Langford Lane, Kidlington, Oxford OX5 1GB, United Kingdom
125 London Wall, London, EC2Y 5AS, United Kingdom

First edition 2022

ISBN: 978-0-323-99397-5
ISSN: 1874-6047

For information on all Academic Press publications
visit our website at https://www.elsevier.com/books-and-journals

Publisher: Zoe Kruze
Acquisitions Editor: Leticia Lima
Developmental Editor: Leticia Lima
Production Project Manager: Abdulla Sait
Cover Designer: Christian Bilbow

Typeset by STRAIVE, India

Contents

Contributors

Iwaki Akiyama
Faculty of Life and Medical Science, Doshisha University, Kyotanabe, Japan

Susumu Fujiwara
Faculty of Materials Science and Engineering, Kyoto Institute of Technology, Kyoto, Japan

Yuji Hatano
Hydrogen Isotope Research Center, University of Toyama, Toyama, Japan

Yuya Higashi
Institute for Integrated Cell-Material Sciences, Institute for Advanced Study, Kyoto University, Kyoto, Japan

Shiori Iida
Genome Dynamics Laboratory, National Institute of Genetics, Research Organization of Information and Systems (ROIS); Department of Genetics, School of Life Science, SOKENDAI (Graduate University for Advanced Studies), Shizuoka, Japan

Tetsuya Kawachi
Kansai Photon Science Institute, Quantum Beam Science Research Directorate, National Institute for Quantum Science and Technology, Kizu, Japan

Takahiro Kenmotsu
Faculty of Life and Medical Sciences, Doshisha University, Kyoto, Japan

Natsuko Kondo
Particle Radiation Oncology Research Center, Institute for Integrated Radiation and Nuclear Science, Kyoto University, Osaka, Japan

Yue Ma
Faculty of Life and Medical Science, Doshisha University, Kyotanabe; Institute for Integrated Cell-Material Sciences, Institute for Advanced Study, Kyoto University, Kyoto, Japan; Institute of Biomedical Engineering, Chinese Academy of Medical Sciences and Peking Union Medical College, Tianjin, China

Kazuhiro Maeshima
Genome Dynamics Laboratory, National Institute of Genetics, Research Organization of Information and Systems (ROIS); Department of Genetics, School of Life Science, SOKENDAI (Graduate University for Advanced Studies), Shizuoka, Japan

Kotaro Matsumoto
Institute for Integrated Cell-Material Sciences, Institute for Advanced Study, Kyoto University, Kyoto, Japan

Katsuhiko Minami
Genome Dynamics Laboratory, National Institute of Genetics, Research Organization of Information and Systems (ROIS); Department of Genetics, School of Life Science, SOKENDAI (Graduate University for Advanced Studies), Shizuoka, Japan

Hiroaki Nakamura
National Institute for Fusion Science, Toki; Graduate School of Engineering, Nagoya University, Nagoya, Japan

Yui Obata
Institute for Quantum Life Science, National Institutes for Quantum Science and Technology, Chiba-shi; Graduate School of Science and Engineering, Ibaraki University, Mito, Japan

Seiki Saito
Graduate School of Science and Engineering, Yamagata University, Yamagata, Japan

Hiroyuki Saitoh
Kansai Photon Science Institute, Quantum Beam Science Research Directorate, National Institutes for Quantum Science and Technology, Hyogo, Japan

Atsushi Shibata
Gunma University Initiative for Advanced Research, GIAR, Gunma University, Maebashi, Japan

Ayumi Shiro
Kansai Photon Science Institute, Quantum Beam Science Research Directorate, National Institutes for Quantum Science and Technology, Hyogo, Japan

Fuyuhiko Tamanoi
Institute for Integrated Cell-Material Sciences, Institute for Advanced Study, Kyoto University, Kyoto, Japan

Akinari Yokoya
Institute for Quantum Life Science, National Institutes for Quantum Science and Technology, Chiba-shi; Graduate School of Science and Engineering, Ibaraki University, Mito, Japan

Kenichi Yoshikawa
Faculty of Life and Medical Sciences, Doshisha University; Center for Integrative Medicine and Physics, Institute for Advanced Study, Kyoto University, Kyoto, Japan

Preface

DNA is under continuous assault from environmental conditions and metabolic changes. This results in various DNA damages that include base modifications and DNA-strand breaks. Of particular importance is double-strand breaks that could lead to genomic instability, resulting in cell death if not repaired. Some of the topics discussed include (1) various types of double-strand breaks, (2) conditions leading to double-strand breaks, including radiation events that produce double-strand breaks, and (3) factors that modify double-strand break formation, including DNA structure. In addition, we discuss various methods used to detect DNA double-strand breaks in a quantitative manner. Thus, this book deals with diverse topics, all of which are centered around DNA double-strand breaks.

We could not have put together this volume in a timely manner without the efforts of the authors who wrote their chapters despite their busy schedule. We thank all the authors for their contributions.

In Volume 52, we will continue with the theme of DNA damages and double-strand breaks. In addition, we plan to include chapters that describe topics and contributors of past volumes. The history of the *Enzymes* series goes back to 1950, and since then a large number of topics including DNA enzymes and biochemistry have been published. We have recently started to document the contents of previous volumes. The next volume will include information on some of these old volumes.

Fuyuhiko Tamanoi
Kenichi Yoshikawa
July 27, 2022
Kyoto, Japan

CHAPTER ONE

Overview of DNA damage and double-strand breaks

Fuyuhiko Tamanoi[a,*] and Kenichi Yoshikawa[b]
[a]Institute for Integrated Cell-Material Sciences, Institute for Advanced Study, Kyoto University, Kyoto, Japan
[b]Faculty of Life and Medical Sciences, Doshisha University, Kyoto, Japan
*Corresponding author: e-mail address: tamanoi.fuyuhiko.2c@kyoto-u.ac.jp

Contents

Abstract

DNA is under a variety of assaults. As a result, different damages accumulate on DNA. These include base changes, single-strand breaks and double-strand breaks. In this volume and also briefly in the following volume, we discuss DNA damage and double-strand breaks. In particular, we focus on double-strand breaks. We discuss types of double-strand breaks as well as methods to detect them. We also discuss how DNA breaks are formed.

1. DNA damages involving base modifications

Continuous assaults on DNA result in DNA damages which include base modifications such as oxidative damages, depurination and depyrimidinations, single-strand breaks (SSBs), double-strand breaks (DSBs), O6-methylguanines and cytosine deaminations. Oxidative damages are frequently observed. An estimation reports 10,000 damages per cell per day in humans [1]. The high frequency reflects the metabolic effect that releases compounds such as reactive oxygen species, reactive nitrogen species, and lipid peroxidation products, and others. Steady-state DNA damages in mammalian cells have been characterized [2,3] and Swenberg et al. [2] reported the following number of damages per cell; 30,000 abasic sites,

The Enzymes, Volume 51
ISSN 1874-6047
https://doi.org/10.1016/bs.enz.2022.08.001

3000 N7-(2-hydroxyethyl)guanine, 2400 8-hydroxyguanosine, etc. Lindahl reported more than 20 types of altered bases [3]. UV light, ionizing radiation and genotoxic chemicals also cause DNA damage [4].

2. DNA strand breaks

DNA damage due to environmental, metabolic factors, as well as spontaneous decay, lead to breaks in DNA. This causes single-strand breaks, a common type of damage arising at a frequency of tens of thousands per cell per day [5]. These breaks are rapidly and efficiently repaired by the SSBR (single-strand repair process). A number of repair mechanisms have evolved and are conserved during evolution pointing to the importance of this type of lesion. Double-strand breaks (DSBs) arise during the attempted repair of damaged DNA bases occurring on opposing strands. In addition, DSBs occur when DNA replication is blocked or paused. DSBs are routinely generated during the treatment of anti-cancer drugs such as cisplatin or phleomycin. Ionizing radiation results in DSBs as well as SSBs [6]. Among the above DNA damages, double-strand breaks (DSBs) are of major concern in life events. If DSBs are not repaired, genomic instability resulting in cell death could occur.

A special type of DNA double-strand breaks is "clustered DNA damage" that occurs, for example, by radiation exposure. This is discovered in 2000 and is defined as a type of lesion that occurs within a small region of the DNA, for example, 10–20 base pair apart [7]. This type of clustered DNA damage contributes significantly to the mutations and genomic instability [8]. Proton beam and carbon ion beam have been used as a way to more efficiently target radiation energy to the tumor. Therapeutic development has been significant with the development of facilities that can provide these cancer cares. The occurrence of clustered DNA breaks by the irradiation with these proton beams and carbon ion beams has been investigated.

In this volume, we discuss DNA double-strand breaks caused by the Auger electrons (see chapters "Core level ionization or excitation and Auger relaxation induce clustered DNA damage" by Yokoya and Obata and "Auger electrons and DNA double-strand breaks studied by using iodine-containing chemicals" by Higashi et al.), α-rays (see chapter "DNA damage and biological responses induced by Boron Neutron Capture Therapy (BNCT)" by Kondo), β-rays (see chapter "Damages of DNA in tritiated water" by Hatano et al.) and heavy ions (see chapter "Carbon ion radiation and clustered DNA double-strand breaks" by Shibata). Mechanical stress such as ultrasound causes DNA double-strand breaks and this topic is discussed in chapter "Mechanical force induced DNA double-strand breaks: Ultrasound" by Ma and Akiyama.

3. Assays to detect DNA double-strand breaks

Various methods to detect DSBs have been developed. This goes back to the 1960s when agarose gel electrophoresis was used to separate different DNA conformations [9,10]. This led to the development of improved methods such as Comet Assay and Halo assay [11]. The Halo assay uses Propidium Iodide that has the property to relax supercoiled DNA and this method has been improved to detect DNA damage at a single-cell level (Fast Halo assay) [11].

An important advance in the detection of DNA double-strand breaks concerns analyses at a single-molecule level. Real-time observation on a single DNA chain has been reported through the application of high-resolution fluorescence microscopy (FM). It is shown that FM provides a quantitative evaluation of the frequency of double-strand breaks for genome-sized DNA molecules in a physiological solution environment. The use of atomic force microscopy (AFM) to measure the length of DNA fragments is also carried out for the specimens adsorbed on solid substrates. These topics are discussed in chapter "Quantitative evaluation of DNA double-strand breaks (DSBs) through single-molecule observation" by Yoshikawa.

A strategy to use DNA damage sensing proteins that are recruited to DSBs has been developed. These proteins include Rad51 involved in forming radiation-induced foci in mammalian cells [12]. Rogakou et al. [13] found that DNA double-strand breaks induce histone H2AX phosphorylation on serine 139. Antibodies recognizing phosphorylated form of histone H2AX have been developed which led to a widespread use of the γH2AX assay [14,15]. Phosphorylation of ATM protein has also been used. Other proteins used that are recruited to DSBs include 53BP1, MRE11 or NSB1.

4. Effect of DNA structure, nucleosomes, and reactive oxygen species

Generation of DNA double-strand breaks is strongly influenced by the structure of DNA such as supercoiling and condensation/compaction. In a study investigating the effect of the Auger electrons in the generation of double-strand breaks, strong influence of supercoiled DNA was observed. Likewise, it is expected that various states of DNA structure will influence the DNA damage including generation of double-strand breaks.

In addition, nucleosome structure present on the DNA influences occurrence of DNA double-strand breaks. DNA is wrapped with core histone octamer that consists of histone proteins (H2A, H2B, H3 and H4). Studies on the effect of nucleosomes on double-strand breaks uncovered major effect of nucleosomes contributing to the protection of DNA from insults such as radiation. The mechanism of this protection is being addressed. These topics are discussed in chapter "Chromatin organization and DNA damage" by Minami et al.

Another important issue when discussing double-strand breaks concerns direct effect and indirect effect on DNA. Reactive oxygen species are generated by insults to cells. In fact, radiation exposure generates reactive oxygen species due to the interaction of X-rays with water molecules (70% of the weight of cells is contributed by water molecules). These reactive oxygen species cause indirect damage to DNA leading to double-strand breaks.

5. Repair of DNA double-strand breaks

Cells have evolved a variety of mechanisms to deal with double-strand breaks. Two different mechanisms are used [16]. One is NHEJ (non-homologous end-joining) which involves the binding of a heterodimer of Ku86 (Ku80)/Ku70 proteins to the ends of the broken DNA. This leads to the binding of DNA-dependent protein kinase. Ligation of the two ends will be accomplished by the complex of XRCC4, XLF (XRCC4 like factor) and DNA ligase IV. Another repair mechanism is called HR (homologous recombination). In this case, single-strand DNA is generated by the action of Mre11/Nbs1/Rad50 complex, CtIP and ExoI/BLM complex. The resulting single-strand DNA region is stably maintained by the binding of RPA (replication protein A). RPA is replaced with Rad51 by the action of BRCA1, BRCA2 (FANCD1), and the single-strand DNA region is covered by the filament of Rad51. DNA synthesis catalyzed by DNA polymerase δ and PCNA is converted to double-strand DNA. Repair is accomplished by further action of Mus81-Eme1, XPF-ERCC1, SLX1-SLX4. Volume 51 is more focused on the detection, as well as the mechanism of double-strand break formation and we will discuss molecular mechanisms of repair in a future volume.

References

[1] B.N. Ames, M.K. Shigenaga, T.M. Hagen, et al., Oxidants, antioxidants, and the degenerative diseases of the aging, Proc. Natl. Acad. Sci. USA 90 (1993) 7915.

[2] J.A. Swenberg, et al., Endogenous versus exogenous DNA adducts: their role in carcinogenesis, epidemiology, and risk assessment, Toxicol. Sci. 120 (Suppl 1) (2011) S130–S145.

[3] T. Lindahl, Instability and decay of the primary structure of DNA, Nature 362 (1993) 709.

[4] A Sancar, et al., Molecular mechanisms of mammalian DNA repair and the DNA damage checkpoints, Annu. Rev. Biochem. (2004), https://doi.org/10.1146/annurev.biochem.73.011303.073723.

[5] K.W. Caldecott, Single-strand break repair and genetic disease, Nat. Rev. Genet. 9 (2008) 619.

[6] J.F. Ward, et al., The complexity of DNA damage: relevance to biological consequences, Int. J. Radiat. Biol. 66 (1994) 427.

[7] B.M. Sutherland, et al., Clustered DNA damages induced in isolated DNA and in human cells by low doses of ionizing radiation, Proc. Natl. Acad. Sci. USA 97 (2000) 103–108.

[8] Z. Nikitaki, et al., In situ detection of complex DNA damage using microscopy: a rough road ahead, Cancers 12 (2020) 3288.

[9] C. Aaij, P. Borst, The gel electrophoresis of DNA, Biochim. Biophys. Acta 269 (1972) 192–200.

[10] J. Vinograd, P Lebowitz, P Laipis, et al., The twisted circular form of polyoma viral DNA, Proc. Natl. Acad. Sci. USA 53 (1965) 1104.

[11] P. Sestili, et al., The fast -Halo assay for the assessment of DNA damage at the single-cell level, Methods Mol. Biol. 521 (2009) 517.

[12] T. Haaf, et al., Nuclear foci of mammalian Rad51 recombination protein in somatic cells after DNA damage and its localization in synaptonemal complexes, Proc. Natl. Acad. Sci. USA 92 (1995) 2298.

[13] E.P. Rogakou, et al., DNA double-strand breaks induce histone H2AX phosphorylation on serine 139, J. Biol. Chem. 273 (1998) 5858.

[14] A. Ivashkevich, et al., Use of the γ-H2AX assay to monitor DNA damage and repair in translational cancer research, Cancer Lett. (2012), https://doi.org/10.1016/j.canlet.2011.12.025.

[15] F.K. Noubissi, et al., Detection and quantification of γ-H2AX using a dissociation enhanced lanthanide fluorescence immunoassay, Sci. Rep. (2012), https://doi.org/10.1038/s41598-021-88296-3.

[16] M. Takata, et al., Homologous recombination and non-homologous end-joining pathways of DNA double-strand break repair have overlapping roles in the maintenance of chromosomal integrity in vertebrate cells, EMBO J. 17 (1998) 5497–5508.

CHAPTER TWO

Quantitative evaluation of DNA double-strand breaks (DSBs) through single-molecule observation

Kenichi Yoshikawa[a,b,*]
[a]Faculty of Life and Medical Sciences, Doshisha University, Kyoto, Japan
[b]Center for Integrative Medicine and Physics, Institute for Advanced Study, Kyoto University, Kyoto, Japan
*Corresponding author: e-mail address: keyoshik@mail.doshisha.ac.jp

Contents

Abstract

By adapting the method of single molecular observation for individual DNAs, it will be shown that reliable analysis of double-strand breaks, DSBs, becomes possible for various kinds of damage sources. Single DNA above the size of several-tens kilo base-pairs exhibits the length scale above several μm, indicating that their whole conformation is visible with fluorescence microscopy by adding suitable fluoresce dye to the solution. Various examples of the quantitative evaluation on DSBs are described, together with the evaluation of the protective effects of anti-oxidants.

1. Introduction

Number of studies have been carried out to detect double-strand breaks, DSBs, of DNA [1]. PCR can be applied to detect DNA damage through the termination of amplification. It is well known that immunological

The Enzymes, Volume 51
ISSN 1874-6047
https://doi.org/10.1016/bs.enz.2022.08.002

assays are useful to detect oxidative DNA damage by using antibody [2–4]. The terminal deoxynucleotidyl transferase (TUNEL) assay has been used to observe DNA fragmentation and strand breaks [5,6]. Experimental methodology for genome-wide DSB mapping has been also reported under the terminology of BLESS (breaks labeling, enrichment on streptavidin and sequencing) [7]. For the detection of DSBs, the comet assay has been applied rather frequently [8–10]. Although these experimental methods have been utilized in many studies, it has been rather difficult to evaluate the number of DBSs at single DNA molecule level, especially for long genomic DNA above the size of hundreds kilo base pairs (kpb) [11]. Recently, it is getting clear that the direct visualization of individual single DNA molecules by fluorescence microscopy provides reliable/useful quantitative information in the analysis of DSBs in giant DNA [12–26]. In the present chapter, various applications of single DNA observation by fluorescence microscopy to detect DSBs in genome-sized DNAs are described.

2. Single-molecule observation on DNA DSBs

Generally, genomic DNA working inside living cells exhibit the size on the order of Mpb–100 Mbp, corresponding to the length scale of the order of mm–10 cm. On the other hand, the diameter of double-stranded DNA is around 2 nm. It is known that double-stranded DNA is characterized as rather stiff double-helical structure, where the persistence length is around 50 nm, corresponding to 150–200 bp. Thus, it is noted that long genomic DNAs behaves as semi-flexible polymer chain, whereas short oligomeric DNAs exhibit the characteristics as stiff rods [27,28]. As the main target of the study on DNA DSBs is long genome-sized DNA molecules, it is essentially important to perform the measurement and evaluation on the generation of DSBs for giant DNAs. As is schematically depicted in Fig. 1, long DNAs above the size of several tens kbp perform folding transition between elongated coil state and folded compact state, accompanied by the change of the solution condition; for example, addition of polyamine [27,28]. On the other hand, short oligomeric DNAs exhibit aggregation without the generation of the folding transition. Here, we will use the term of folding transition for the conformational change on single DNA molecules, to avoid confusion between the single chain event and multi-chain phenomenon such as aggregation and precipitation. The fundamental characteristics of DNAs depending on their size have been clarified/confirmed based on the single molecule observation by fluoresce microscopy [27,28].

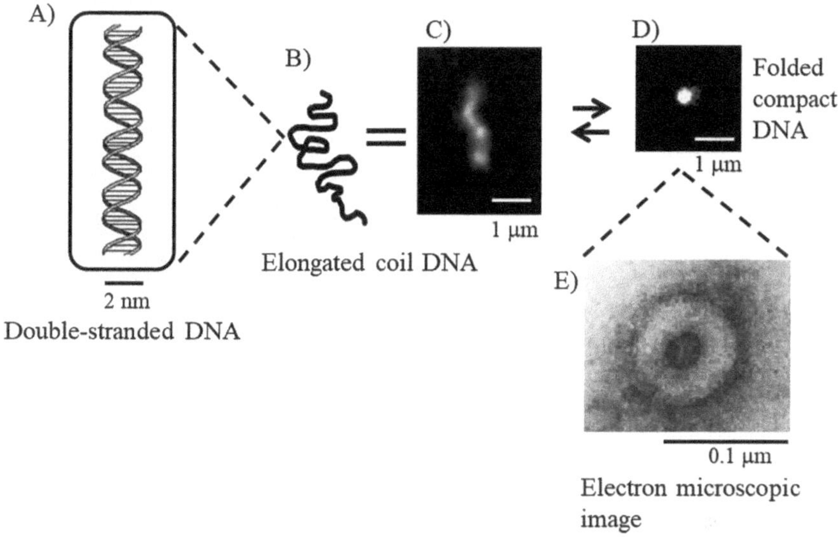

Fig. 1 (A, B) Long DNA above the size of several tens kbp behaves as a semiflexible chain, whereas short oligomeric DNA (below the size of several hundred bp) is characterized as rigid rod. (C, D) Long DNA undergoes the transition of higher-order structure from elongated coil onto folded compact state, as depicted as the fluorescence microscopic images on the elongate and compact states, respectively, for the T4 GT7 DNA (166 kbp). The compact state is generated by the addition of spermidine (3+). (E) is the electron microscopic image of the folded compact DNA. It is noted that, owe to the blurring effect with the fluorescence microscopic measurement, the compact DNA in (D) looks larger than its actual size exemplified in (E).

In this chapter, we may show that single molecular observation is a potent, useful experimental method to evaluate DSBs in a quantitative manner and also to get useful information on the mechanism of DNA breakage, for example to distinguish whether the DSB is caused through one-step or two-step.

3. DSBs caused by γ-ray

Fig. 2 exemplifies the fluorescence microscopic images of T4 GT7 DNA (166 kbp) fixed on glass surface after the exposure to ^{60}Co γ-ray irradiation, where DNAs are stained by fluorescence dye, YOYO-1 [19].

For the evaluation of the probability of DSBs, average contour length $L(I)$ of approximately 100 molecular chains at each does I (Gy) was measured. Then, the number of DSBs per base pair, P_0, was calculated

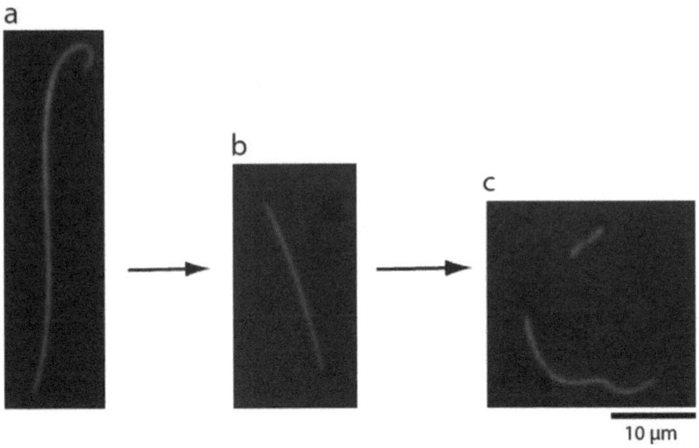

Fig. 2 Examples of fluorescence images for the DNA molecules after γ-ray irradiation for T4 GT7 DNA. Fluorescence dye YOYO-1 was used for the visualization. DNA molecules in aqueous solution were stretched on the surface through the effect of a weak shear caused by a mild slip of glass slides and then fixed on the glass surface pretreated with poly-L-lysine solution. (a) 0 Gy, (b) 50 Gy, and (c) 100 Gy. *Adapted from S.F. Shimobayashi, T. Iwaki, T. Mori, K. Yoshikawa, Probability of double-strand breaks in genome-sized DNA by γ-ray decreases markedly as the DNA concentration increases, J. Chem. Phys. 138 (2013) 174907/1-5.*

from the data of the average DNA lengths of irradiated, $L(I)$, and non-irradiated control, $L(0)$, samples.

$$P_0 = \frac{L(0)/L(I) - 1}{L(0)} \cdot \eta \tag{1}$$

where, η (= 0.34 nm) is the distance between the adjacent base pair in double-stranded DNA.

Fig. 3(A) shows the change of the probability of DSBs, P_0, as the function of γ-ray dose, I, at different DNA concentrations from 0.015 to 300 μM in base pair units. It is found that P_0 is linearly dependent on I for all of the concentrations of DNA measured. From the slope on the linear dependence of P_0 vs I given in Fig. 3(A), probability of DSBs per Gy (adsorbed dose of the γ-ray irradiation), P_1, is calculated. Fig. 3(B) shows log-log plot of P_1 vs C_{bp} (base pair concentration of DNA). Inspection of the graph indicates the existence of the two different regimes. For low concentration region, P_1 remains almost constant irrespective to C_{bp}. Contrary to this, for higher concentrations, P_1 tends to decrease with the increase of C_{bp}, being almost proportional to C_{bp}^{-1}. In other words, the total number of DSBs remains essentially constant on this region, indicating that the major

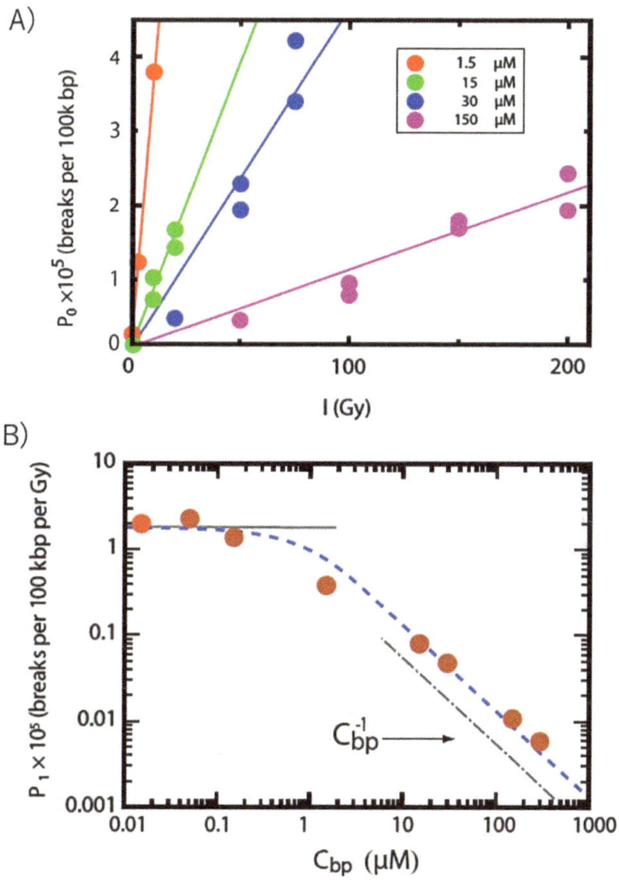

Fig. 3 (A) Number of DSBs per base pair, P_0, as a function of the γ-ray irradiation dose from ^{60}Co source. The concentration indicated by different colors are base-pair concentration, C_{bp}, of T4 GT7 DNA. (B) Log-log plot of P_1 on the number of DSBs per 100 kilo base pair (kbp) and per Gy as a function of DNA base-pair concentration, C_{bp}. *Adapted from S.F. Shimobayashi, T. Iwaki, T. Mori, K. Yoshikawa, Probability of double-strand breaks in genome-sized DNA by γ-ray decreases markedly as the DNA concentration increases, J. Chem. Phys. 138 (2013) 174907/1-5.*

mechanism of DSBs is indirect through the attack by reactive species generated by the photon of γ-ray. Such conclusion on the mechanism of DSBs corresponds well to the study by Ito et al. [29]; the contribution of indirect effects to DSBs at $C_{bp} \approx 100$ μM is approximately two orders of magnitude larger than that of direct effect. The experimental data given in Fig. 3 reveals that the size of DNA and its characteristics as semiflexible polymers play an essential role on the efficiency of DNA damage cause by ionization radiation with γ-ray.

4. Real-time observation of DSBs

Direct observation of single giant DNA molecules in solution provides a useful methodology of real time measurement on the DSBs. In Fig. 4(A) is exemplified [13] the real-time observation of breakage of a single T4 GT7 DNA in an aqueous solution containing Fenton's reagent (Fe^{2+}/H_2O_2)

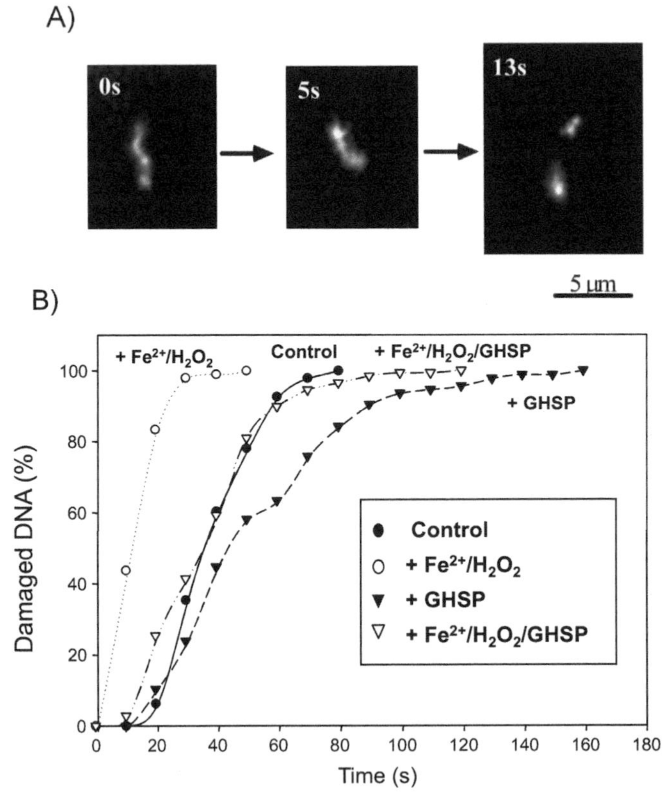

Fig. 4 (A) Real-time observation on DSB for a single T4 GT7 DNA molecule stained by YOYO-1 through the observation by fluorescence microscopy. Active oxygen was generated by the Fenton reaction with 12 mM H_2O_2 and 2.5 μM Fe^{2+}. (B) Time-dependence of the percentage of damaged DNA molecules evaluated from the real-time observation of DSB. *Adapted from Y. Yoshikawa, M. Suzuki, N. Yamada, K. Yoshikawa, Double-strand break of giant DNA: protection by glucosyl-hesperidin as evidenced through direct observation on individual DNA molecules, FEBS Lett. 566 (2004) 39–42.*

[30]. Before the breakage, single DNA exhibits translational and intrachain fluctuation, as the Brownian motion. After the moment of the breakage, pair of fragments tend to eliminate each other gradually under Brownian motion. Fig. 4(B) shows the time course for the increase of damaged DNA with double-strand breaks, which have been evaluated from the direct measurement of the DSBs as exemplified in Fig. 4(A). For the control experiment, [Control], the mean breakage time is around 37 sec after the start of the single molecular observation by fluorescence microscopy. Such breakage in the control experiment is attributable to the chemical attack by hydroxyl radials generated through photo-irradiation on YOYO-1 (the chemical formula is shown in the upper of Fig. 5, which was added to the solution to visualized DNA molecules under fluorescence microscopy. Actually, in these observations, we made the UV illumination with the wavelength of $\lambda = 450$–$490\,nm$. It is known that YOYO-1 has the ability to cause photo-breakage of DNAs by generating ROSs, including hydroxyl radicals

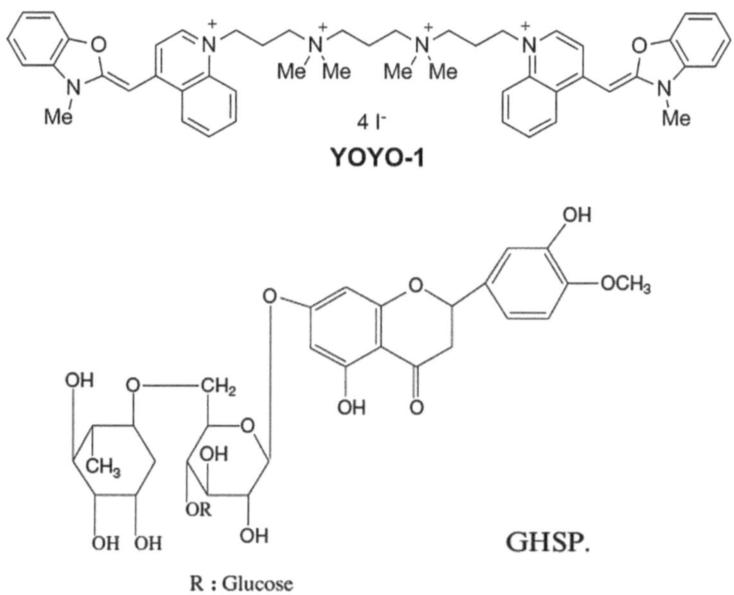

Fig. 5 Chemical Formulas of YOYO-1 and GHSP (glucosyl-hesperidin). *The formula of GHSP is adapted from Y. Yoshikawa, M. Suzuki, N. Yamada, K. Yoshikawa, Double-strand break of giant DNA: protection by glucosyl-hesperidin as evidenced through direct observation on individual DNA molecules, FEBS Lett. 566 (2004) 39–42.*

[31–33]. The breakage curve indicates that the breakage is accelerated markedly with the addition of Fenton's reagent, by comparing the data of [Control] with that of [+Fe^{2+}/H_2O_2]. Contrary to this, the breakage rate is retarded for the DNA in the presence of an antioxidant, glucosyl-hesperidin (GHSP, the chemical formula is shown in the bottom of Fig. 5. The result shown in Fig. 4(B) indicate that GHSP exhibits the protective effect to reduced the probability of double-strand breaks to the level of [Control] even in the solution environment generating high amount of oxidative species with the Fenton reaction.

As the next, we may discuss the mechanism of the double-strand breaks under UV irradiation on DNA in the presence of YOYO-1. It was suggested that YOYO-1 causes DSBs of DNA through the accumulation of single-strand breaks [31]. Thus, it is expected that, during UV irradiation, single-strand breaks will be induced in a stochastic manner along the long double-stranded DNA. As the next step, DSBs along long DNA will be generated when chemical attach is made near a nick, or the site of the single-strand break [see the scheme given in Fig. 6(A)]. Based on such hypothesis on the mechanism of DSBs, we will derive a kinetic equation to interpret the time-dependent profile of the reaction of DSBs.

By denoting the number of nicks per single DNA molecule as n, the rate of the increase in nicks (step I of the scheme given in Fig. 6(A) is given as,

$$dn/dt = \alpha I \tag{2}$$

where I is the intensity of illuminated light, or the strength of photon flux, and α is a constant. For simplicity, we assume that variation in the base composition along the DNA chain has a negligible effect on the possibility of breakage. As the initial condition, we take $n=0$ when $t=0$. Thus, Eq. (2) is integrated as,

$$n = \alpha It \tag{3}$$

The reaction kinetics of step II in the scheme of Fig. 6(A) , i.e., the probability of chemical attack at the nick, is represented as dx/dt, where x is the number of DNA molecules surviving from fragmentation (see, e.g., Fig. 4(B). By introducing a rate constant k, dx/dt is given as,

$$dx/dt = -knx = -k\alpha Itx \tag{4}$$

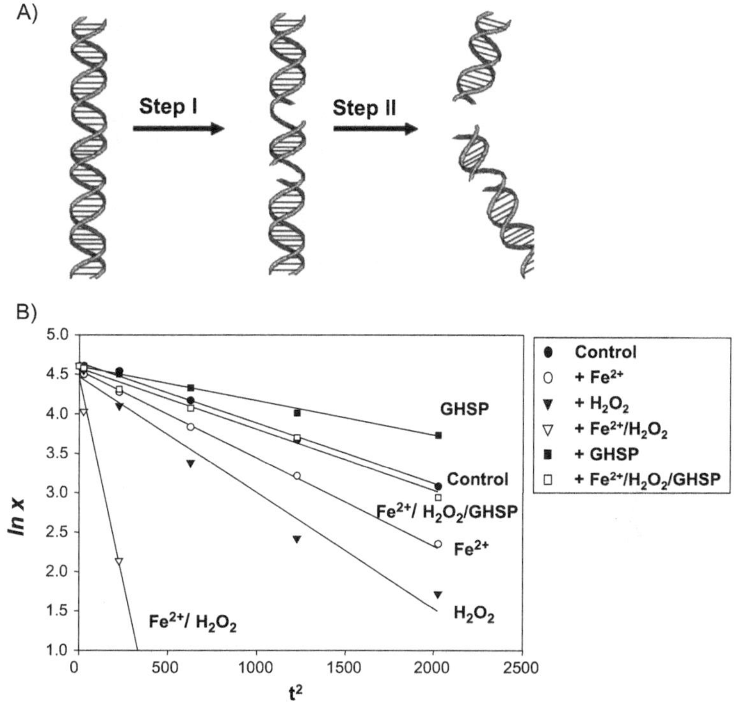

Fig. 6 (A) Scheme on the DSB caused through two-steps reactions of single strand breaks. (B) Relationships between t^2 and ln x, where x is the percentage of surviving DNA molecules and is calculated as 100% - [percentage of damaged DNA]. The linear relationships in the graph suggest the mechanism of DSB through two-steps as in the schematics of (A). Similar experimental procedures of real-time observations were performed as in Fig. 4(A). *Adapted from Y. Yoshikawa, M. Suzuki, N. Yamada, K. Yoshikawa, Double-strand break of giant DNA: protection by glucosyl-hesperidin as evidenced through direct observation on individual DNA molecules, FEBS Lett. 566 (2004) 39–42.*

After integration of Eq. (4),

$$\ln x = -(1/2)k\alpha It^2 + \text{Const.} \tag{5}$$

By introducing an effective kinetic constant, $A = (1/2)k\alpha I$, Eq. (5) becomes

$$\ln x = -At^2 + \text{Const.} \tag{6}$$

Table 1 Mean breakage time <τ> until double-strand breakage (DSB) and the effective constant A of the breakage reaction, where A is normalized to be unity ($A_0 = 1$) in the control experiment.

Sample	Mean breakage time <τ> (s)	Effective kinetic constant[a] A/A_0
Control (DNA alone)	37	1.0
$+Fe^{2+}$	30	1.3
$+H_2O_2$	25	1.7
$+Fe^{2+} + H_2O_2$	12	12.2
+GHSP	52	0.4
$+Fe^{2+} + H_2O_2$ + GHSP	36	0.9

[a]Experimental error is estimated to be ±0.2.

Thus, when double-strand breakage is induced in a two-step reaction with stochastic single-strand damage and subsequent attack by reactive oxygen, a plot of $\ln x$ vs t^2 should provide a straight line, where the slope of the linear relationship is proportional to the rate constant k of the chemical breakage reaction on long DNA molecules, if the reaction rate α of stochastic photo-breakage remains almost constant.

Fig. 6(B) shows linear relationships between t^2 and $\ln x$, suggesting that the above two-step mechanism of DSB explains the experimental trend in a satisfactory manner. In addition, the figure shows that the effective kinetic constant A changes significantly depending on the presence or absence of GHSP. The kinetic constant A in the experiment with reactive oxygen is 12 times greater than that in the control experiment. In the presence of reactive oxygen with hydrogen peroxide and Fe^{2+}, the kinetic constant A with GHSP is $12.2/0.9 \approx 13$ times smaller than that without GHSP (see Table 1). The protective effect of GHSP described above corresponds well to the reported experimental trends on flavonoids as potential antioxidant [34].

5. DNA compaction exhibits marked protective effect against DSBs

As shown in Fig. 1, single giant DNA larger than several tens kbp undergoes a large discrete transition from an elongated coil state to a folded compact state by the addition of various condensing agents such as multivalent polyamines, solvable polymers (PEG, albumin, etc.), and cationic surfactant [27,28,35–37]. It is noted that, in living cells, giant DNAs are condensed into confinements, for example, genomic DNAs with the total length of 2 m is stored into nuclear with the size of several μm. Thus, it is important to make clear how the probability of DSBs is affected through the compaction/condensation of genome-sized DNA. Here, we may show the application of single DNA observation to evaluate the effect of higher-structural change on the probability of DSBs in a quantitative manner [15]. As described below, we have measured the probability of DSBs caused by the irradiation with Co-60 γ-ray, for the elongated and compact states under the same concentration of spermidine, SPD(3+).

Fig. 7(A) shows fluorescence microscopic images of individual T4 GT7 DNA molecules observed in bulk solution. In buffer solution (10 mM Tris-HCl buffer, pH 7.4), individual DNA molecules exist in an elongated coil state (Fig. 7(A)a) [15]. Upon treatment with 0.8 mM spermidine (3+), elongated DNA was folded into a compact state (Fig. 7(A)b). Such compact DNA was unfolded into an elongated state by the addition of 300 mM NaCl (final concentration) (Fig. 7(A)c). It has been found that such unfolding phenomenon ofDNA is caused by the addition of 1:1 salt, because of the competitive effect of the translational entropy and charge neutralization [38]. In the lower part in Fig. 7(A) is shown the fluorescence microscopic images for the DNA molecules fixed upon a glass surface under weak shear of the sample solutions, where a', b' and c' are the DNA samples corresponding to a, b, c in bulk solution, respectively. We have pretreated the glass surface with the absorbance of cationic poly-(L-lysine) solution. The change of the contour length, or full length, of individual DNA is measured directly from the fluorescence images for the stretched state. Fig. 7(B) shows the experimental scheme to compare the effect of same amount of spermidine for the DSBs under irradiation of cobalt-60 γ-ray. Similar to the procedure in Section 3 in the present chapter, we may analyze the probability ofDSBs on the γ-ray dose. We may consider that the number of DSBs, n, is proportional to the dose D of γ-ray.

$$L(\mathrm{I}) \approx L(0)/(n+1) \tag{7}$$

It may be assumed that the number of DSBs, n, is proportional to the dose I of the γ-ray.

$$n = \mathrm{k}I \tag{8}$$

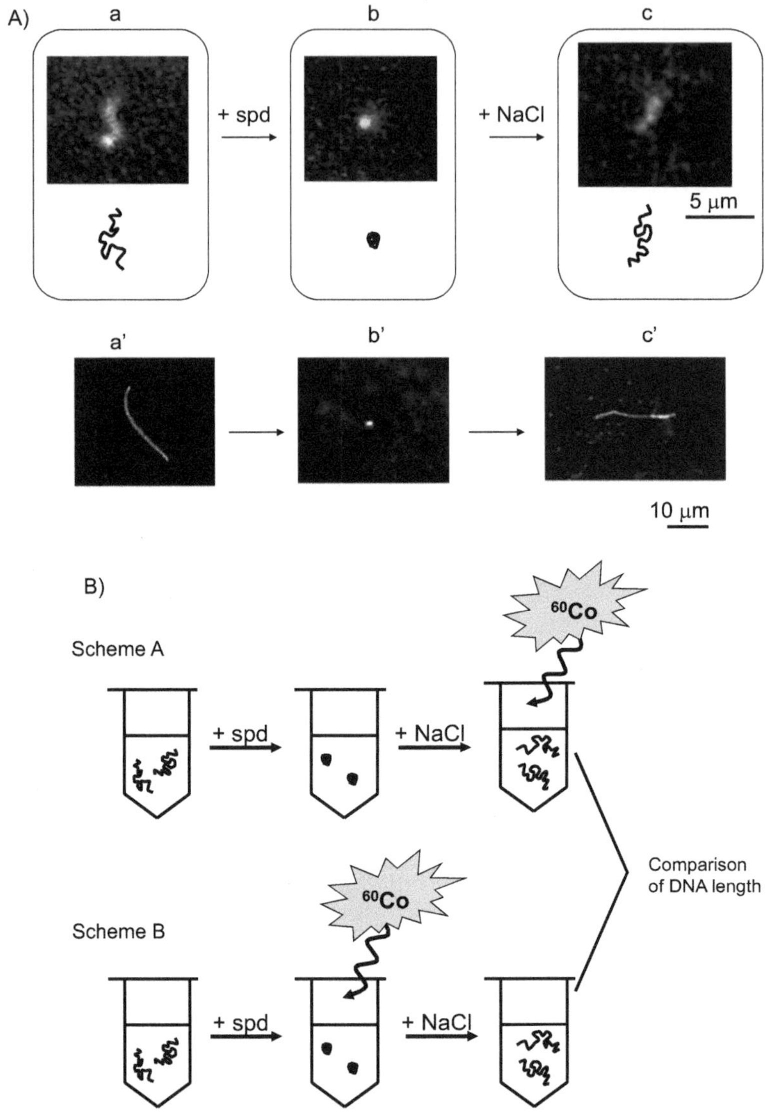

Fig. 7 (A) Fluorescence microscopic images of T4 GT7 DNA exhibiting Brownian fluctuation in bulk solution (above) and fixed on a glass surface under weak shear (below). (a, a′) Elongated coil conformation in the absence of both spermidine (SPD) and NaCl. (b, b′) Folded compact conformation in the presence of 0.8 mM SPD. (c, c′) Elongated coil conformation in the presence of both 0.8 mM SPD and 300 mM NaCl. (B) Experimental procedure for the observation of protective effect of SPD by comparing the different state of DNA with elongated and compact states. *Adapted from Y. Yoshikawa, T. Mori, N. Magome, K. Hibino, K. Yoshikawa, DNA compaction plays a key role in radioprotection against double-strand breaks as revealed by single-molecule observation, Chem. Phys. Lett. 456 (2008) 80–83.*

where k is a proportional constant reflecting the probability of DSBs. From Eqs. (7) and (8), the following relation is obtained.

$$L(I)/L(0) = 1/(\mathrm{k}I + 1) \tag{9}$$

From the fitting cure base on Eq. (9), the radiation efficiency k's of DSBs are obtained (the detail of the analysis is described in Ref. [15]).

Fig. 8 shows the dependency of the efficiency k on the concentration of spermidine (SPD) [15]. This indicates that, in the presence of 0.8 mM SPD, compact DNA (scheme A) is about 25 times less susceptible to the elongated coil DNA (scheme B). In other words, the protective effect of SPD almost negligible small for the elongated state (scheme A). This implies that the direct radical-scavenging effect of SPD is almost negligible, at least under the experimental condition adapted in this study.

It has, thus, become evident that the decrease in the solvent exposure of DNA in the compact state is the main cause for the significant decrease of susceptibility to the γ-ray irradiation.

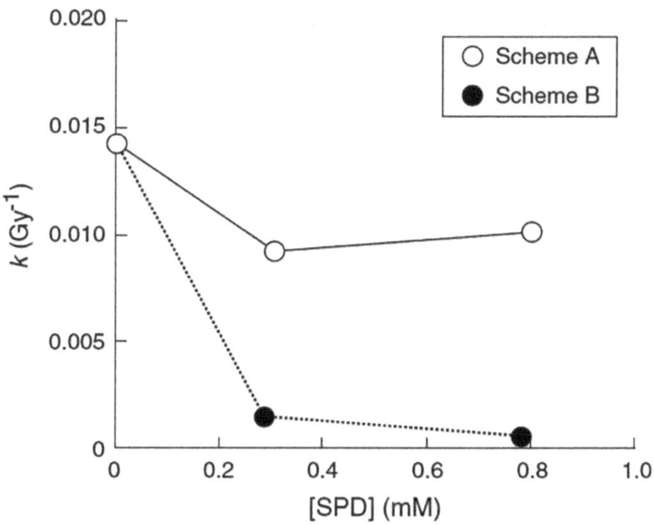

Fig. 8 Efficiency of DSBs by γ-ray irradiation, k, vs spermidine, SPD(3+) concentration. *Adapted from Y. Yoshikawa, T. Mori, N. Magome, K. Hibino, K. Yoshikawa, DNA compaction plays a key role in radioprotection against double-strand breaks as revealed by single-molecule observation, Chem. Phys. Lett. 456 (2008) 80–83.*

6. How to reduce DSBs caused by mechanical mixing

It is well known that giant DNA molecules are easily broken into fragments by mechanical stress, such as mixing and pipetting, during the usual experimental procedures used to isolate and purify them from the in-cell genome. On the other hand, it is getting clear that higher-order structural change of genomic DNA concerns with the underlying mechanism of epigenetic phenomena. Thus, it is important to develop the experimental technique to treat genome-sized DNA molecules by avoiding the occurrence of DSBs, especially for the indispensable procedure of mechanical mixing for aqueous samples in the experiments of molecular biology and biochemistry. In this section, we may show the results of single molecule observation on the frequency of DSBs after mechanical mixing. Based on the experimental results, a useful method of sample mixing to reduced DSBs in a marked manner.

Fig. 9(A) shows the experimental apparatus for the mechanical mixing of an aqueous solution of DNA. A microtube containing 300 μL of DNA solution was fixed to the center of a Vortex mixer (MS3, Thermo Fisher Scientific, USA). The experimental results indicated that the probability of DSBs per 10 kbp is larger for T4 GT7 DNA (166 kbp) than for λ DNA (49 kbp) by 1.2–1.3 times, suggesting that larger DNA exhibits greater damage probability caused by mechanical stress. Single molecular observations on DSBs by mechanical mixing were performed to compare the different manners of mixing between conventional and proposed procedures: Constant rotation from the start of mixing (indicated by green color) and gradual increase of the rotation rate (indicated by red color), respectively, as schematically shown on the time-dependency of the rotation rate f_R as in Fig. 9(B). For the experiment with conventional procedure, the number of DSBs increases significantly above the level of 0.5, per 10 kbp, almost abruptly for the initial period (1–2 s from the start of mixing). As for the proposed procedure, the speed or rotation was gradually increased from 0 to 1300 rpm over 60 s. Even after the attainment to the speed of 1300 rpm, the degree of DSBs remain in a relatively low level. Thus, it has become clear the mixing procedure proposed here suppresses DSBs by >50% compared to constant strength of mixing. The significantly high probability of DSBs at the start of mixing in the conventional procedure is attributable to the breaks caused near the wall boundary, which is

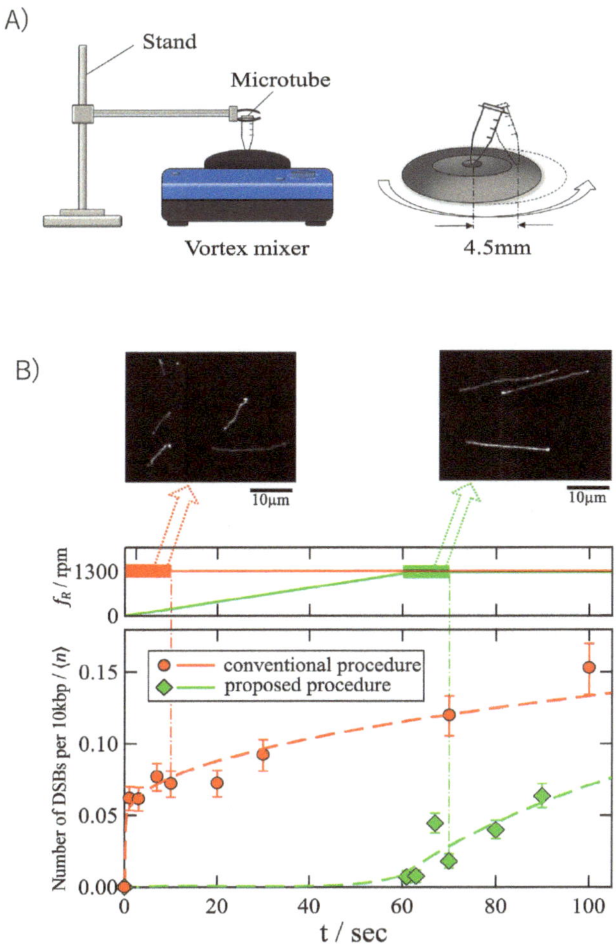

Fig. 9 Measurement of DSBs caused by mechanical mixing. (A) Experimental apparatus for mixing. The bottom of microtube was rotated with an orbit diameter of 4.5 mm. (B) Difference in the degree of DSBs between mixing with different rpm profiles (upper panel); conventional procedure with a constant mixing strength (red), and the proposed procedure with a gradual increase in the intensity of mixing (green). Upper pictures: Fluorescence microscopic images of DNA molecules after 10 s mixing at 1300 rpm exemplified for the both procedures. *Adapted from H. Kikuchi, K. Nose, Y. Yoshikawa, K. Yoshikawa, Double-strand breaks in genome-sized DNA caused by mechanical stress under mixing: quantitative evaluation through single-molecule observation, Chem. Phys. Lett. 701 (2018) 81–85.*

the location of the greatest shear stress under rotational mechanical agitation (for example; see the well-established theoretical model of the Hagen-Poiseuille flow [39]). Within a few second from the start of mixing, most of the DNA molecules may migrate toward the inner part of the vessel due to the large shear stress near the vessel wall, even under low speed of mixing. Thus, it is essentially important to avoid the sudden stress for the mixing procedure in order to avoid the damage of DSBs by mechanical mixing.

7. Protective effect of antioxidant against various damage sources

In the preceding section, experimental results on the quantitatively evaluate the number of DSBs through direct DNA observations have been described. Here, we will show the application of single DNA observation for the analysis of protective effect of antioxidant to different damage sources; photo-irradiation, γ-ray-irradiation and freezing. As the antioxidant, dimethyl sulfoxide (DMSO) is adapted. DMSO is known to be a useful free radical scavenger and a radio-protectant [40–45]. It has been reported that DMSO decreases the degree of radiation injury of neighboring organs in cancer radio-therapy. It is also noted that DMSO is one of the most important agents in cryopreservation [40,41,46,47], where DMSO is regarded to have the role to protect against DSBs owe to its strong solvation effect on water molecules.

Fig. 10 shows the relative kinetic constant, k, on the occurrence of DSBs at different injury sources, evaluated based on methodology of the single molecular measurement [22]. As the injury sources to cause DSBs, photo-induced reactive oxygen, γ-ray irradiation and freeze/thawing were adapted. To observe the effect of photo-induced reactive oxygen (ROSs; reactive oxygen species), YOYO-1 was used as a photosensitizer to generate ORS, and the real-time observation of DSBs for individual DNA molecules was performed, where YOYO-1 also exhibits the role to visualize DNA under visible light. With respect to the damage by γ-ray irradiation, the number of DSBs was evaluated based on the experimentally observed average length of DNA molecules at different degree of irradiation. As for the damage caused by freezing/thawing, the frequency of DSBS was measured with two different cold temperatures, −25 °C and −80 °C. Fig. 10 indicates the DSBs caused by photo- and γ-ray-irradiation markedly decrease to the order of 1/100 when the concentration of DMSO is above 1.5–2.0%.

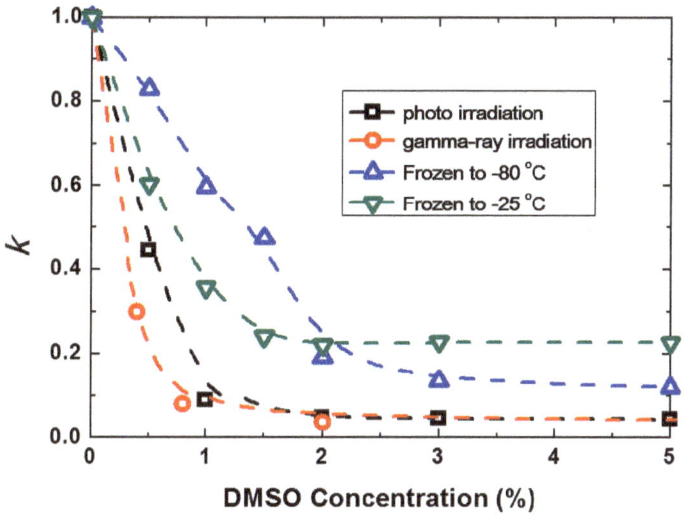

Fig. 10 Difference in the protective effect of DMSO. Vertical axis is the relative kinetic constant $k = K/K_0$ for the generation of DSBs at different concentrations of DMSO, where K_0 is the rate constant in the absence of DMSO. With respect to freezing, "quick freezing" is to −80 °C, and "slow freezing" is to −25 °C. *Adapted from M. Noda, Y. Ma, Y. Yoshikawa, T. Imanaka, T. Mori, M. Furuta, T. Tsuruyama, K. Yoshikawa, A single-molecule assessment of the protective effect of DMSO against DNA double-strand breaks induced by photo-and g-ray-irradiation, and freezing, Sci. Rep. 7 (2017) 8557/1-8.*

On the other hand, the decrease in the rate constant k of DSBs remains in the order of 1/5–1/10 for the damage due to freeze/thawing, even at a DMSO concentration above 2%. For all of the injuries studied here, the protective effects of DMSO almost maximized around 2% and keep stationary value at higher DMSO concentrations. This would be a useful knowledge, because past studies reported very toxic biological effect at high concentrations of DMSO [45,48].

8. Future prospective

Currently, it has been considered that a single double-strand break, DSB, is very serious with the possibility to kill a cell if it inactivates a key gene. Unfortunately, it has been rather difficult to measure DSBs, especially at very low damage conditions. In the present chapter, it is shown that kinetics on the fragmentation process of giant DNA can be evaluated through the methodology of single molecular observation with fluorescence microscopy, on the order of one DSB per 100 kbp. It is highly expected that such methodology of the quantitative evaluation of DSBs through single molecule observation would be applied actively in a rich variety fields of biological and medical sciences. For example, application of real time observation on DSBs for the heavy ion radiotherapy would provide reliable data toward the improvement of efficiency with larger safety.

In addition to the measurement of DSBs, the methodology of single molecular observation would be very useful to uncover the role of higher-order structural change on genetic control. Although the images obtained by single molecular observation through optical microscopy provide relatively low resolution in contrast to EM, electron microscopy, and AFM, atomic force microscopy, single molecule observation affords dynamic images with fluctuation, translational and intramolecular Brownian motion in aqueous solution environment. Actually, recent studies with single molecular observation have been demonstrating the importance on the understanding of dynamical behavior of the higher order structure in genome-sized DNAs [49–58] and in chromatin [59,60]. Further researches utilizing single molecular observation are expected to help us to gain insight on the essential problems in life science, i.e., spatio-temporal dynamical behavior and characteristics of living systems.

Acknowledgments

This work was partially supported by JSPS KAKENHI Grant Number JP20H01877 and JP20H05934. The author thanks all of the collaborators as shown for the citations in the References.

References

[1] S. Kumari, R.P. Rastogi, K.L. Singh, S.P. Singh, R.P. Sinha, DNA damage: detection strategies, EXCLI J. 7 (2008) 44–62.

[2] A.M. Amaro, K.B. Hallberg, E.B. Lindstrom, C.A. Jerez, An immunological assay for detection and enumeration of thermophilic biomining microorganisms, Appl. Environ. Microbiol. 60 (1994) 3470–3473.

[3] L.J. Kuo, L.-X. Yang, γ-H2AX - a novel biomarker for DNA double-strand breaks, In Vivo 22 (2008) 305–309.

[4] X. Liu, F. Li, Q. Huang, Z. Zhang, L. Zhou, Y. Deng, M. Zhou, D.E. Fleenor, H. Wang, M.B. Kastan, C.-Y. Li, Self-inflicted DNA double-strand breaks sustain tumorigenicity and stemness of cancer cells, Cell Res. 27 (2017) 764–785.

[5] Y. Gavrieli, Y. Sherman, S.A. Ben-Sasson, Identification of programmed cell death in situ via specific labeling of nuclear DNA fragmentation, J. Cell Biol. 119 (1992) 493–501.

[6] H. Takata, T. Hanafusa, T. Mori, M. Shimura, Y. Iida, K. Ishikawa, K. Yoshikawa, Y. Yoshikawa, K. Maeshima, Chromatin compaction protects genomic DNA from radiation damage, PLoS One 8 (2013) e75622/1–11.

[7] B.A.M. Bouwman, F. Agostini, S. Garnerone, G. Petrosino, H.J. Gothe, S. Sayols, A.E. Moor, S. Itzkovitz, M. Bienko, V. Roukos, N. Crosetto, Genome-wide detection of DNA double-strand breaks by in-suspension BLISS, Nat. Protocols 15 (2020) 3894–39418.
[8] A.R. Collins, The comet assay for DNA damage and repair: principles, applications, and limitations, Mol. Biotechnol. 26 (2004) 249–261.
[9] W. Liao, M.A. McNutt, W.G. Zhu, The comet assay: a sensitive method for detecting DNA damage in individual cells, Methods 48 (2009) 46–53.
[10] P. Sykora, K.L. Witt, P. Revanna, S.L. Smith-Roe, J. Dismukes, D.G. Lloyd, B.P. Engelward, R.W. Sobol, Next generation high throughput DNA damage detection platform for genotoxic compound screening, Sci. Rep. 8 (2018) 2771/1–20.
[11] W.J. Cannan, D.S. Pederson, Mechanism and consequence of double-strand DNA break formation in chromatin, J. Cell. Physiol. 231 (2016) 3–14.
[12] Y. Yoshikawa, T. Kanbe, K. Yoshikawa, Daunomycin triggers membrane blebbing and breakage of giant DNA encapsulated in a cell-sized liposome, Chem. Phys. Lett. 366 (2002) 305–310.
[13] Y. Yoshikawa, M. Suzuki, N. Yamada, K. Yoshikawa, Double-strand break of giant DNA: protection by glucosyl-hesperidin as evidenced through direct observation on individual DNA molecules, FEBS Lett. 566 (2004) 39–42.
[14] Y. Yoshikawa, K. Hizume, Y. Oda, K. Takeyasu, S. Araki, K. Yoshikawa, Protective effect of vitamin C against double-strand breaks in reconstituted chromatin visualized by single-molecule observation, Biophys. J. 90 (2006) 993–999.
[15] Y. Yoshikawa, T. Mori, N. Magome, K. Hibino, K. Yoshikawa, DNA compaction plays a key role in radioprotection against double-strand breaks as revealed by single-molecule observation, Chem. Phys. Lett. 456 (2008) 80–83.
[16] M. Suzuki, C. Crozatier, K. Yoshikawa, T. Mori, Y. Yoshikawa, Protamine-induced DNA compaction but not aggregation shows effective radioprotection against double-strand breaks, Chem. Phys. Lett. 480 (2009) 113–117.
[17] Y. Yoshikawa, T. Mori, M. Suzuki, T. Imanaka, K. Yoshikawa, Comparative study of kinetics on DNA double-strand break induced by photo- and gamma-irradiation: Protective effect of water-soluble flavonoids, Chem. Phys. Lett. 501 (2010) 146–151.
[18] K. Yoshida, N. Ogawa, K. Yukihiro, H. Tabata, Y. Watanabe, T. Kenmotsu, Y. Yoshikawa, K. Yoshikawa, Effect of low-frequency ultrasound on double-strand breaks in giant DNA molecules, Appl. Phys. Lett. 103 (2013) 063705/1–3.
[19] S.F. Shimobayashi, T. Iwaki, T. Mori, K. Yoshikawa, Probability of double-strand breaks in genome-sized DNA by γ-ray decreases markedly as the DNA concentration increases, J. Chem. Phys. 138 (2013) 174907/1–5.
[20] Y. Ma, N. Ogawa, Y. Yoshikawa, T. Mori, T. Imanaka, Y. Watanabe, K. Yoshikawa, Protective effect of ascorbic acid against double-strand breaks in giant DNA: Marked differences among the damage induced by photo-irradiation, gamma-rays and ultrasound, Chem. Phys. Lett. 638 (2015) 205–209.
[21] R. Kubota, Y. Yamashita, T. Kenmotsu, Y. Yoshikawa, K. Yoshida, Y. Watanabe, T. Imanaka, K. Yoshikawa, Double-strand breaks in genome-sized DNA caused by ultrasound, ChemPhysChem 18 (2017) 959–964.
[22] M. Noda, Y. Ma, Y. Yoshikawa, T. Imanaka, T. Mori, M. Furuta, T. Tsuruyama, K. Yoshikawa, A single-molecule assessment of the protective effect of DMSO against DNA double-strand breaks induced by photo-and g-ray-irradiation, and freezing, Sci. Rep. 7 (2017) 8557/1–8.
[23] H. Kikuchi, K. Nose, Y. Yoshikawa, K. Yoshikawa, Double-strand breaks in genome-sized DNA caused by mechanical stress under mixing: quantitative evaluation through single-molecule observation, Chem. Phys. Lett. 701 (2018) 81–85.

[24] Y. Hatano, Y. Konaka, H. Shimoyachi, T. Kenmotsu, Y. Oya, H. Nakamura, Kinetics of double strand breaks of DNA in tritiated water evaluated using single molecule observation method, Fusion Eng. Design 146 (2019) 100–102.
[25] Y. Ma, K. Ishihara, K. Yoshida, I. Akiyama, K. Yoshikawa, Double-strand breaks in genome-sized DNA Caused by Megahertz Ultrasound, J. Acoust. Soc. Am. 150 (2021) 241–247.
[26] S. Saito, H. Nakamura, T. Kenmotsu, Y. Oya, Y. Hatano, Image processing method for automatic measurement of number of DNA breaks, J. Adv. Simul. Sci. Eng. 8 (2021) 173–193.
[27] K. Yoshikawa, Y. Yoshikawa, Compaction and condensation of DNA, in: R.I. Mahato, S.W. Kim (Eds.), Pharmaceutical Perspectives of Nucleic Acid-based therapy, Taylor & Francis, London, 2002, pp. 136–163.
[28] A. Estevez-Torres, D. Baigl, DNA compaction: fundamentals and applications, Soft Matter 7 (2011) 6746–6756.
[29] T. Ito, S.C. Baker, C.D. Stickley, J.G. Peak, M.J. Peak, Dependence of the yield of strand breaks induced by γ-rays in DNA on the physical conditions of exposure: water content and temperature, Int. J. Radiat. Biol. 426 (1993) 289–296.
[30] D.R. Lloyd, D.H. Phillips, Oxidative DNA damage mediated by copper(II), iron(II) and nickel(II) Fenton reactions: evidence for site-specific mechanisms in the formation of double-strand breaks, 8-hydroxydeoxyguanosine and putative intrastrand cross-links, Mutation Res. 424 (1999) 23–36.
[31] B. Akerman, E. Tuite, E. Single- and Double-Strand Photocleavage of DNA by YO, YOYO and TOTO, Nucleic Acids Res. 24 (1996) 1080–1090.
[32] C. Kanony, B. Akerman, E. Tuite, Photobleaching of asymmetric cyanines used for fluorescence imaging of single DNA molecules, J. Am. Chem. Soc. 123 (2001) 7985–7995.
[33] J.R. Pyle, K.W.E. Sy Piecco, J.R. Vicente, J. Chen, In situ sensing of reactive oxygen species on dye-stained single DNA molecules under illumination, Langmuir 35 (2019) 11308–11314.
[34] K. Ohtsuki, A. Abe, H. Mitsuzumi, M. Kondo, K. Uemura, Y. Iwasaki, Y. Kondo, Effects of long-term administration of hesperidin and glucosyl hesperidin to spontaneously hypertensive rats, J. Nutr. Sci. Vitaminol. 48 (2002) 420–422.
[35] S.M. Mel'nikov, V.G. Sergeyev, K. Yoshikawa, Discrete coil-globule transition of large DNA induced by cationic surfactant, J. Am. Chem. Soc. 117 (1995) 2401–2408.
[36] V.V. Vasilevskaya, A.R. Khokhlov, Y. Matsuzawa, K. Yoshikawa, Collapse of single DNA molecule in poly(ethylene glycol) solutions, J. Chem. Phys. 102 (1995) 6595–6602.
[37] K. Yoshikawa, M. Takahashi, V.V. Vasilevskaya, A.R. Khokhlov, Large discrete transition in a single DNA molecule appears continuous in the ensemble, Phys. Rev. Lett. 76 (1996) 3029–3031.
[38] K. Hibino, Y. Yoshikawa, S. Murata, T. Saito, A.A. Zinchenko, K. Yoshikawa, Na^+ more strongly inhibits DNA compaction by spermidine (3+) than K^+, Chem. Phys. Lett. 426 (2006) 405–409.
[39] S.P. Sutera, R. Skalak, The history of Poiseuille's law, Annu. Rev. Fluid Mech. 25 (1993) 1–19.
[40] S.W. Jacob, J.C. De La Torre, Dimethyl Sulfoxide (DMSO) in Trauma and Disease, CRC Press, Taylor & Francis Group, 2015.
[41] Z.W. Yu, P.J. Quinn, Dimethyl sulfoxide: a review of its applications in cell biology, Biosci. Rep. 14 (1994) 259–281.
[42] Y.-Y. Hsiao, T.-H. Hung, Computational evaluation of biological effects of dimethylsulfoxide for radiotherapy, Chung Shan. Med. J. 23 (2012) (2012) 107–116.

[43] R. Runge, L. Oehme, J. Kotzerke, R. Freudenberg, The effect of dimethyl sulfoxide on the induction of DNA strand breaks in plasmid DNA and colony formation of PC Cl3 mammalian cells by alpha-, beta-, and Auger electron emitters (223)Ra, (188)Re, and (99 m)Tc, EJNMMI Res. 6 (2016) 1–10.
[44] J.E. Repine, O.W. Pfenninger, D.W. Talmage, E.M. Berger, D.E. Pettijohn, Dimethyl sulfoxide prevents DNA nicking mediated by ionizing radiation or iron/hydrogen peroxide-generated hydroxyl radical, Proc. Natl. Acad. Sci. U. S. A. 78 (1981) 1001–1003.
[45] A. Bishayee, D.V. Rao, L.G. Bouchet, W.E. Bolch, R.W. Howell, Protection by DMSO against cell death caused by intracellularly localized iodine-125, iodine-131 and polonium-210, Radiat. Res. 153 (2000) 416–427.
[46] J.E. Lovelock, M.W. Bishop, Prevention of freezing damage to living cells by dimethyl sulphoxide, Nature 183 (1959) 1394–1395.
[47] A.M. Schrader, C.-Y. Cheng, J.N. Israelachvili, S. Han, Contrasting effects of glycerol and DMSO on lipid membrane surface hydration dynamics and forces, J. Chem. Phys. 145 (2016) 041101/1–4.
[48] D. Berz, E.M. McCormack, E.S. Winer, G.A. Colvin, P.J. Quesenberry, Cryopreservation of hematopoietic stem cells, Am. J. Hematol 82 (2007) 463–472.
[49] Y.T. Sato, S.N. Watanabe, T. Kenmotsu, M. Ichikawa, Y. Yoshikawa, J. Teramoto, T. Imanaka, A. Ishihama, K. Yoshikawa, Structural change of DNA induced by nucleoid proteins: growth phase-specific Fis and stationary phase-specific Dps, Biophys. J. 105 (2013) 1037–1044.
[50] Y. Shimizu, Y. Yoshikawa, T. Kenmotsu, S. Komeda, K. Yoshikawa, Conformational transition of DNA by dinuclear Pt(II) complexes causes cooperative inhibition of gene expression, Chem. Phys. Lett. 678 (2017) 123–129.
[51] A. Kanemura, Y. Yoshikawa, W. Fukuda, K. Tsumoto, T. Kenmotsu, K. Yoshikawa, Opposite effect of polyamines on In vitro gene expression: enhancement at low concentrations but inhibition at high concentrations, PLoS One 13 (2018) e0193595/1–11.
[52] T. Kishimoto, Y. Yoshikawa, K. Yoshikawa, S. Komeda, Different effects of cisplatin and transplatin on the higher-order structure of DNA and gene expression, Inter. J. Mol. Sci. 21 (2020) 34/1–15.
[53] H. Tanaka, C.-Y. Shew, Y. Yoshikawa, T. Kenmotsu, K. Yoshikaw, Low-efficiency of gene expression with a long diamine is attributable to the effect on DNA zipping, Chem. Phys. Lett. 745 (2020) 137253–137259.
[54] H. Sakuta, F. Fujita, T. Hamada, M. Hayashi, K. Takiguchi, K. Tsumoto, K. Yoshikawa, Self-emergent protocells generated in an aqueous solution with binary macromolecules through liquid-liquid phase separation, ChemBioChem 21 (2020) 3323–3328.
[55] T. Nishio, K. Sugino, Y. Yoshikawa, M. Matsumoto, Y. Oe, K. Sadakane, K. Yoshikawa, K^+ promotes the favorable effect of polyamine on gene expression better than Na^+, PLoS One 15 (2020) e0238447/1–13.
[56] T. Kitagwa, T. Nishio, Y. Yoshikawa, N. Umezawa, T. Higuchi, C.-Y. Shew, T. Kenmotsu, K. Yoshikawa, Effects of structural isomers of spermine on the higher-order structure of DNA and gene expression, Int. J. Mol. Sci 22 (2021) 2355–2368.
[57] T. Nishio, Y. Yoshikawa, K. Yoshikawa, S. Sato, Longer DNA exhibits greater potential for cell-free gene expression, Sci. Rep. 11 (2021) 11739/1–7.
[58] T. Nishio, Y. Yoshikawa, K. Yoshikawa, Higher-order structure of DNA determines its positioning in cell-size droplets under crowded conditions, PLoS One 16 (2021) e0261736/1–13.
[59] S.S. Ashwin, K. Maeshima, M. Sasai, Heterogeneous fluid-like movements of chromatin and their implications to transcription, Biophys. Rev. 12 (2020) 461–468.
[60] T. Yamamoto, T. Sakaue, H. Schiessel, Slow chromatin dynamics enhances promoter accessibility to transcriptional condensates, Nucleic Acids Res. 49 (2021) 5017–5027.

CHAPTER THREE

Chromatin organization and DNA damage

Katsuhiko Minami[a,b], Shiori Iida[a,b], and Kazuhiro Maeshima[a,b,*]
[a]Genome Dynamics Laboratory, National Institute of Genetics, Research Organization of Information and Systems (ROIS), Shizuoka, Japan
[b]Department of Genetics, School of Life Science, SOKENDAI (Graduate University for Advanced Studies), Shizuoka, Japan
*Corresponding author: e-mail address: kmaeshim@nig.ac.jp

Contents

The Enzymes, Volume 51
ISSN 1874-6047
https://doi.org/10.1016/bs.enz.2022.08.003

Abstract

Genomic DNA is organized three-dimensionally in the nucleus as chromatin. Recent accumulating evidence has demonstrated that chromatin organizes into numerous dynamic domains in higher eukaryotic cells, which act as functional units of the genome. These compacted domains facilitate DNA replication and gene regulation. Undamaged chromatin is critical for healthy cells to function and divide. However, the cellular genome is constantly threatened by many sources of DNA damage (e.g., radiation). How do cells maintain their genome integrity when subjected to DNA damage? This chapter describes how the compact state of chromatin safeguards the genome from radiation damage and chemical attacks. Together with recent genomics data, our finding suggests that DNA compaction, such as chromatin domain formation, plays a critical role in maintaining genome integrity. But does the formation of such domains limit DNA accessibility inside the domain and hinder the recruitment of repair machinery to the damaged site(s) during DNA repair? To approach this issue, we first describe a sensitive imaging method to detect changes in chromatin states in living cells (single-nucleosome imaging/tracking). We then use this method to explain how cells can overcome potential recruiting difficulties; cells can decompact chromatin domains following DNA damage and temporarily increase chromatin motion (~DNA accessibility) to perform efficient DNA repair. We also speculate on how chromatin compaction affects DNA damage-resistance in the clinical setting.

1. Introduction

1.1 DNA, histones, and nucleosomes

A nucleosome is formed when negatively charged DNA is wrapped around a core histone octamer, which is comprised of two copies of four positively charged histone proteins (H2A, H2B, H3 and H4) [1–3]. Each histone octamer has 1.7 left-handed superhelical turns composed of 147 base pairs of DNA [4]. Nucleosome particles are connected by linker DNA segments that generate repetitive motifs of ~200 bp and are depicted as "beads on a string" or as a "10-nm fiber." Basic core histones neutralize about only half of the DNA negative charges derived from the phosphate backbone, resulting in electrostatic repulsion between adjacent regions of this 10-nm fiber [5]. Intrinsically disordered regions (IDRs) are found on amino-terminal tail domains that extend towards the exterior of the nucleosome from each histone. H2A also contains an extra carboxy-terminal tail domain. Therefore, 10 tails containing 10 IDRs are present on each nucleosome [4,6] and they are highly positively charged due to a preponderance of lysines and arginines. Due to their abundant positive charges, these long, disordered tail domains facilitate nucleosome interactions that enable them to fold.

Other factors required to screen the remaining negative charges to further fold the 10-nm fiber include linker histones, nonhistone proteins, cations and other positively charged molecules [7].

1.2 Chromatin condensation *in vitro*

As mentioned above, the type and concentration of ions in the surrounding environment can heavily influence local chromatin structure. Cations like magnesium ion (Mg^{2+}) can facilitate interactions between nucleosomes, as shown in the formation of cation (Mg^{2+})-dependent chromatin condensates *in vitro*, which requires nucleosome–nucleosome interactions by histone tails (Fig. 1A; [15–19]: for reviews, see [7,21]). In pure water or in the absence of Mg^{2+}, chromatin is extended to resemble "beads on a string" due to the electrostatic repulsion between adjacent nucleosomes (Fig. 1A, left). Mg^{2+} counteracts the net negative charge and repulsion by weakly associating with chromatin DNA phosphate backbones (Fig. 1A, center). Increasing the concentration of Mg^{2+} can result in an almost complete neutralization of this repulsion and enable interdigitated nucleosome–nucleosome interactions. Histone tails conduct these interactions and enable irregular chromatin folding to form condensates (Fig. 1A, right).

1.3 Compact chromatin domains in the cell

In addition to the *in vitro* results, emerging evidence has demonstrated that nucleosomes form irregularly folded clusters/condensates within the nucleus of the cell. In higher eukaryotes, light and electron microscopy have revealed chromatin organization with a diameter of ~200 nm [22–25], while open chromatin arrangements predominate in yeast cells where only a few nucleosomes may cluster together [26–28]. Recently, "chromatin domains" have been discovered using a combination of 3D super-resolution and scanning EM techniques. These chromatin domains appeared as variably sized groups of nucleosomes (~200 nm) [29,30]. Of note, the chromatin domains in euchromatic regions maintain compact chromatin organizations with active epigenetic marks, but their sizes are smaller than those of heterochromatin, which is consistent with the previous reports [23,31]. Electron spectroscopic imaging (ESI) is a technique with high contrast and resolution, which allows the visualization of 10-nm fibers by mapping phosphorus and nitrogen atoms [32]. ESI showed similar chromatin domain structures [18]. Note that concentrations of Mg^{2+} and other

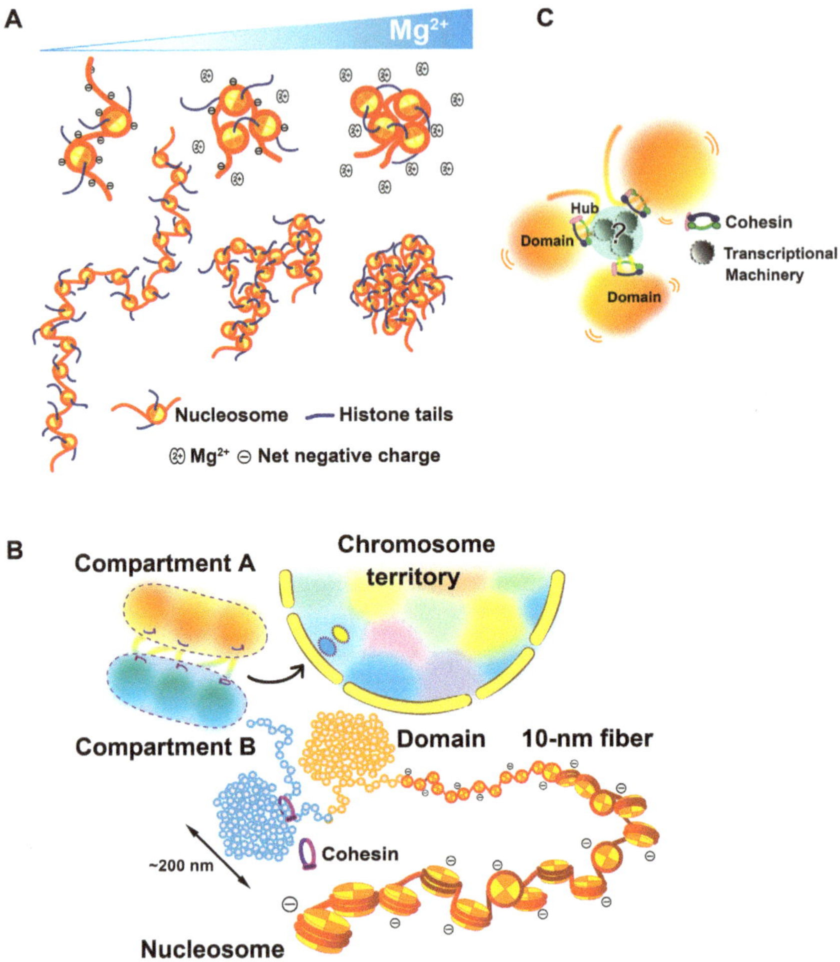

Fig. 1 Nucleosome folding and hierarchy of chromatin organization. (A) Schematic of cation-dependent (Mg^{2+}) compaction of chromatin. Electrostatic repulsion stretches chromatin when cations and other positively charged factors are absent (left). The phosphate backbone of DNA weakly interacts with Mg^{2+} in chromatin (i.e., previously referred to as an ion atmosphere [8]). The net negative charge and repulsion of chromatin decreases as Mg^{2+} levels increase (center), chromatin histone tails begin to dominate the nucleosome–nucleosome repulsions and promote local chromatin folding (center). Nucleosome repulsion is almost eliminated with higher levels of Mg^{2+}, promoting interdigitation of nucleosomes (condensate formation) facilitated through histone tail interactions (right). (B) Simplified schematic of nuclear chromatin organization. The 10-nm fiber is compacted into chromatin domains (e.g., topologically associating domain [TAD]/contact domain/loop domain) [9-12,20], which interact over long distances to form chromatin compartments [13]. Compartments generally represent a transcriptionally active or open chromatin state (compartment A) and an inactive or closed chromatin state (compartment B). A single interphase chromosome is occupied in a chromosome territory (highlighted as different colors) [14]. (C) Chromatin forms numerous compact domains as functional units of the genome and compaction/decompaction of the domains can regulate a variety of genome functions. *Work was reproduced with permission from the authors for panel (A) Y. Itoh, S. Iida, S. Tamura, R. Nagashima, K. Shiraki, T. Goto, K. Hibino, S. Ide, K. Maeshima, 1,6-Hexanediol rapidly immobilizes and condenses chromatin in living human cells, Life Sci. Alliance, 4 (2021) e202001005 and panels (B and C) K. Maeshima, S. Iida, S. Tamura, Physical nature of chromatin in the nucleus, Cold Spring Harb. Perspect. Biol. 13 (2021) a040675.*

positively charged factors under physiological conditions seem to be sufficient to form irregular condensates, as seen *in vitro*.

Recent 3C/Hi-C methods have uncovered the hierarchy organization of the genome. This technology is used to map genomic DNA interactions by identifying the probability of fine contacts [9,33] and has unveiled how the genome in higher eukaryotes is subdivided into chromatin domains (Fig. 1B). These domains consist of topologically associating domains (TADs) [10–12] and smaller contact/loop domains (average size of ~185kb) [20]. Super-resolution imaging visually confirmed these structures [34]. Cohesin appears to maintain the organization of loop domains (Fig. 1B) [34–37]. For more details, see [33]). Chromatin domains can be further bundled into megabase scaled compartments (compartment A and B) [13], which typically are transcriptionally active or inactive chromatin, respectively (Fig. 1B). There is a collective hierarchy of nuclear chromatin organization (Fig. 1B). 10-nm fibers are organized into domains, domains into compartments A and B, and then into nuclear "chromosome territories" that contain individual chromosomes [14]. The large-scale domain arrangements detailed above appear to be essential for higher level chromatin features in the nucleus [38]. Of note, domains are arranged compactly (Fig. 1B), regardless of being composed of euchromatin or heterochromatin [18,23,30]. Histone tails mediate nucleosome–nucleosome interactions that are essential for domains to compact locally [18,23]. These interactions can also be expediated by cations like Mg^{2+}, as seen *in vitro* by increasing condensation of chromatin as the concentration of Mg^{2+} is increased (Fig. 1A; [15–19]).

1.4 Functional relevance of chromatin domains

The ability of larger protein complexes (e.g., transcription factors and replication initiation complexes) to access DNA target sites may be modulated by the compaction state of various chromatin domains [30,31] because the target DNA sequence of these complexes might be hindered by the compact domains [30,31]. Decompaction of such domains with histone modifications or by the action of other proteins can increase accessibility for these complexes to commence gene transcription [18,23,30,39–41]. Thus, chromatin forms numerous compact domains as functional units of the genome (Fig. 1C), and compaction/decompaction of domains can regulate a variety of genome functions [7].

2. Chromatin compaction and radiation damage

As discussed above, the organization of compact chromatin domains in living cells plays an important role in genome functions including DNA replication and gene regulation. Besides this, are there any other roles for these chromatin domains? Recent studies indicated that the highly compacted nature of chromatin contributes to maintaining genomic integrity, including resistance against radiation damage. We discuss this issue in the following subsections.

2.1 Development of a solid-phase system for handling chromatin and detecting DNA damage

Previous *in vitro* studies demonstrated that DNA compaction helped protect against double-strand breaks (DSBs) generated by γ-rays [42–46]. These findings suggested that higher-order chromatin structures observed in the cell may also contribute to DNA damage protection. Indeed, several cellular studies have demonstrated that foci of serine-139 phosphorylation in histone H2AX (γH2AX), which is a marker for DNA DSBs [47,48], were frequently detected in euchromatin regions and less frequently in supposedly more condensed heterochromatic regions [49–53]. Furthermore, higher levels of γH2AX foci were observed in cells with decondensed chromatin induced by a hypotonic treatment [49]. Although these studies implied that chromatin compaction protected genomic DNA against damage, it should be noted that the cellular results might have been caused by many possible indirect effects [54,55], making simple conclusions difficult to draw.

Our group established an *in vitro* system for sensitive DNA damage detection to directly demonstrate that compacted chromatin plays a crucial role in protecting against DSBs (Fig. 2; [58]). This system uses nuclei isolated from human cells and can clearly quantify DNA damage caused by ionizing irradiation [58].

Previously, there were no efficient *in vitro* systems that could manipulate long chromatin effectively and quantitatively detect generated DSBs because long genomic DNA is easily damaged during experimental manipulations. To prevent artifactual physical damage to our genomic DNA before and during our analyses, we used permeabilized nuclei that were attached to glass surfaces. The nuclei were isolated from HeLa cells and attached to poly-L-lysine-coated coverslips by gentle centrifugation (Fig. 2A).

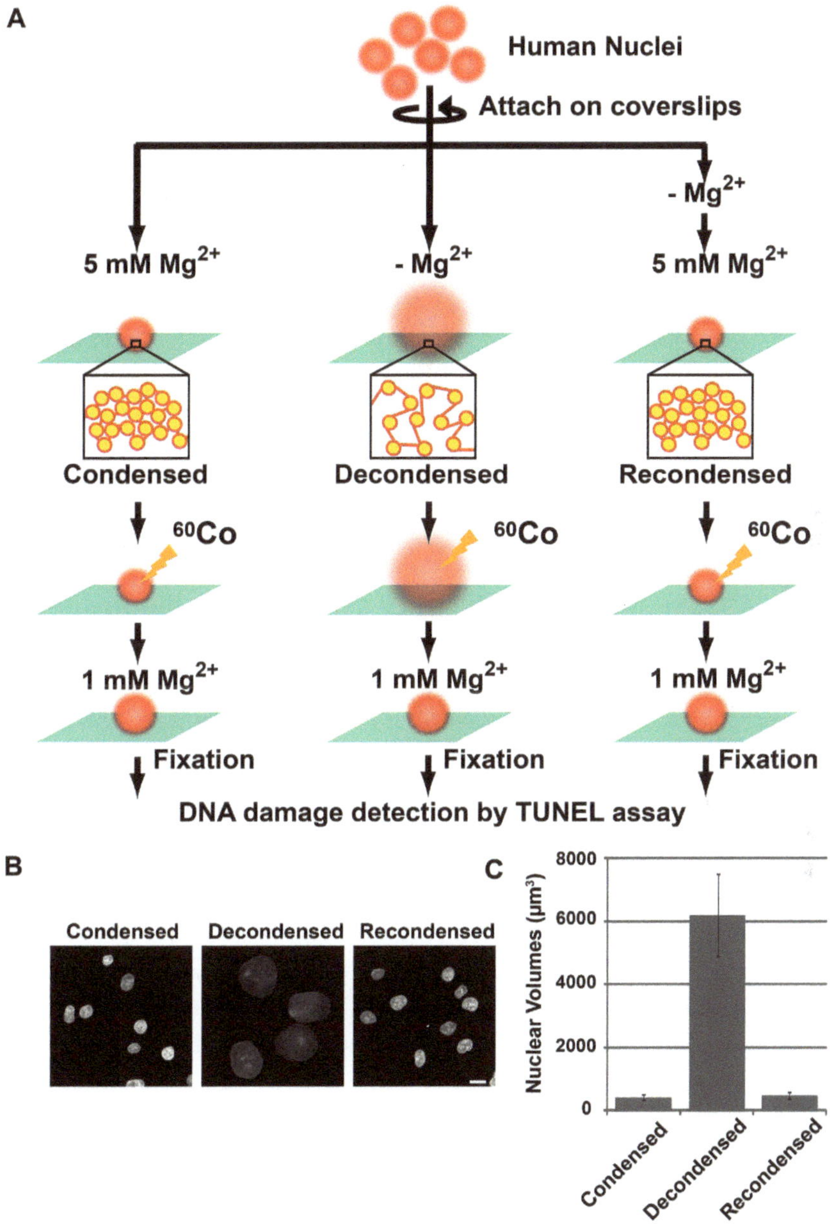

Fig. 2 Illustration of how chromatin is manipulated to detect DNA damage in the "solid-phase" system. (A) Chromatin inside purified nuclei attached to a glass surface were treated in the presence (condensed; left and see Panel B) or absence of 5 mM Mg^{2+} (decondensed; middle and see Panel B). When Mg^{2+} (5 mM) was restored to the decondensed nuclei/chromatin, chromatin was recondensed (right and see Panel B). The presence and absence of Mg^{2+} (5 mM) is sufficient to reconstruct the condensation and decondensation states of chromatin, respectively, as cations can alter the compaction state of chromatin (i.e., phase-transition process [56]). Condensed, decondensed

As discussed above, divalent cations condense chromatin (Fig. 1A). Negatively charged chromatin repulses other chromatin when cations are absent, but will condense in a dose-dependent manner if these negative charges are neutralized by cations like Mg^{2+} (Fig. 1A) [15,59,19]. Thus, we can manipulate the size and compaction of nuclei and chromatin by altering the concentrations of Mg^{2+}. Chromatin became highly condensed at 5 mM Mg^{2+}, whereas it decondensed in the absence of Mg^{2+} [15,56,60] (Fig. 2A, also see Fig. 2B). We estimated the concentration of chromatin based on the nuclear volume (Fig. 2C) and known size of the genomic DNA. This "solid-phase system" allowed us to conduct very gentle and quantitative handling of the genome-sized DNA. Furthermore, we fluorescently labeled DSB sites using terminal deoxynucleotidyl transferase (TdT) (i.e., a TUNEL assay [57]) to directly detect DNA damage without inducing additional breakages in the chromatin adhered to the glass surface (Figs. 2A and 3A).

2.2 DNA damage caused by γ-ray irradiation is diminished in compacted chromatin

We quantitatively investigated whether chromatin compaction in interphase nuclei protects against γ-ray irradiation induced DNA damage. As shown in Fig. 2B, 5 mM Mg^{2+} condensed nuclei and associated chromatin (~400 μm³) on the glass surface. When Mg^{2+} was absent, nuclei and chromatin became decondensed and swelled almost 15-fold (~6000 μm³) (Fig. 2C). To determine if DNA decondensation itself would result in damage, nuclei and chromatin were recondensed (Fig. 2B) using increasing amounts of Mg^{2+} until they were comparable in size to the original condensed nuclei (~450 μm³) (Fig. 2C). We irradiated these three types of nuclei (condensed, decondensed and recondensed) with an escalating range of γ-ray (cobalt-60) exposures (Fig. 2A). Post irradiation, a uniform compact

and recondensed nuclei irradiated with varying doses of cobalt-60 γ-rays were then treated with a buffer containing 1 mM Mg^{2+}. The TUNEL assay was used to detect chromatin DSBs [57] (bottom). Here, minor fractions of single-strand breaks (SSBs) might be also detected. (B) Nuclear images stained for DNA (DAPI) demonstrating the increased size of decondensed nuclei compared to condensed and recondensed nuclei. Bar, 10 μm. (C) Quantification of nuclear volumes for condensed, decondensed and recondensed nuclei. *All data were reproduced and modified with permission from H. Takata, T. Hanafusa, T. Mori, M. Shimura, Y. Iida, K. Ishikawa, K. Yoshikawa, Y. Yoshikawa, K. Maeshima, Chromatin compaction protects genomic DNA from radiation damage, PLoS One 8 (2013) e75622.*

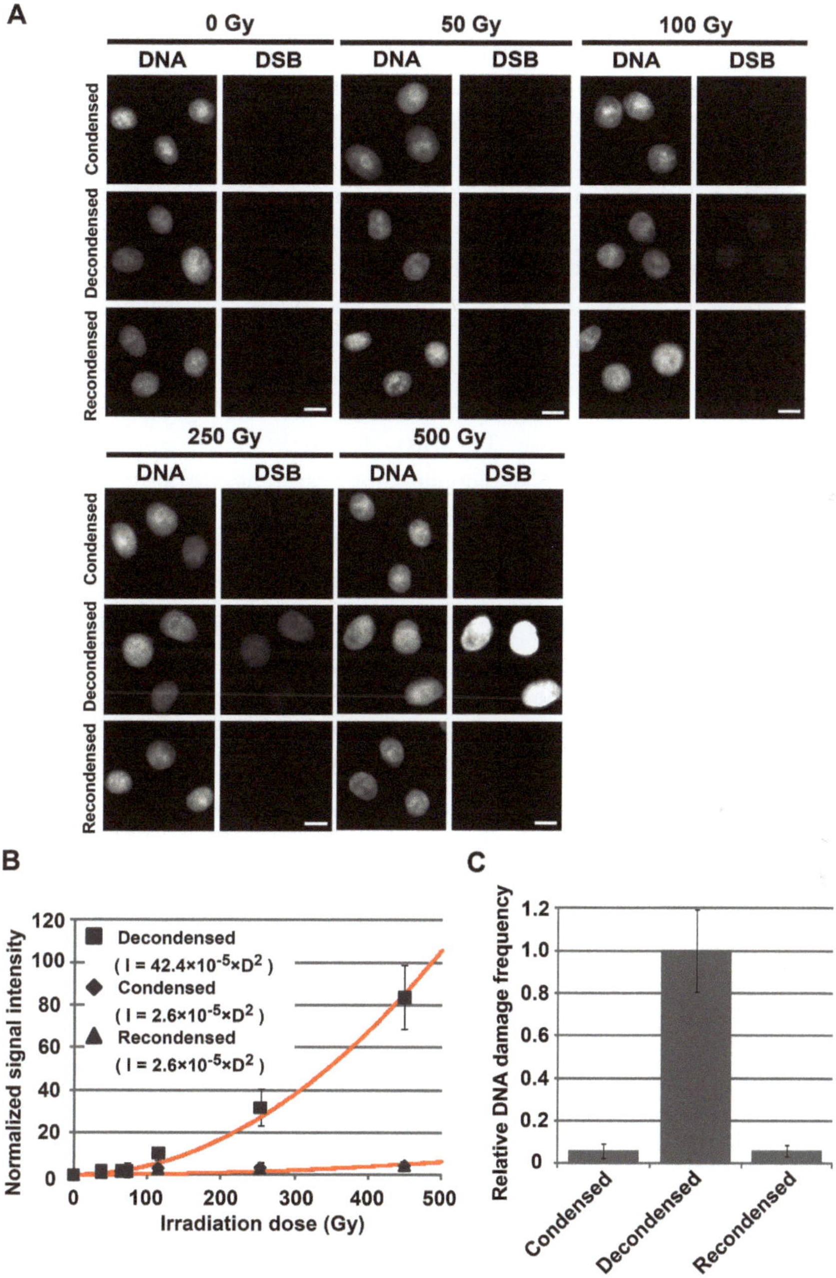

Fig. 3 Compacted chromatin represses γ-ray irradiation-induced DNA damage. (A) The TUNEL assay was used to detect DSBs after treating condensed (1st row), decondensed (2nd row) and recondensed (3rd row) nuclei with increasing doses of γ-ray irradiation and then incubating all 3 types of irradiated nuclei in a buffer containing 1 mM Mg^{2+} to attain similar nuclear sizes for imaging. Uniform nuclear size allowed equal accessibility to the chromatin and similar treatment post irradiation. Images of DNA (left) and DSB

state was induced by treating nuclei with 1 mM Mg^{2+} to ensure chromatin from each group would be equally accessible and handled similarly among the three nuclei types. Nuclei were then fixed with formaldehyde (Fig. 2A). The TUNEL assay was used to detect chromatin DNA damage (Figs. 2A and 3A).

Interestingly, DSB signals from decondensed chromatin strongly escalated in a dose-dependent manner, while only faint signals were observed in the chromatin of condensed and recondensed nuclei (Fig. 3A). The very low signals detected in recondensed chromatin indicated that the decondensation process did not cause DNA damage (Fig. 3A; third row). A quadratic curve ($I = kD^2$) closely fit our plot of DNA damage over an increasing dose of irradiation, where: I = DNA breakage; k = a constant; and D = the irradiation dose (Fig. 3B). When k values were compared, a 16-fold difference in DNA damage suppression was observed between condensed and decondensed chromatin (interphase nuclei; Fig. 3C) [58].

Isolated human mitotic chromosomes were used to further confirm these observations [61]. Like interphase nuclei, mitotic chromosomes expanded almost 50-fold in the absence of Mg^{2+}, compared to compact mitotic chromosomes. The results obtained using chromosomes were more striking than those obtained using nuclei, in that the compact chromosomes had approximately 50-fold greater resistance to γ-ray irradiation than the decondensed chromosomes [58]. The induction of DNA damage highly correlates to the compaction status and subsequent concentration (or volume) of chromatin. Notably, the quadratic curve (Fig. 3B) indicates that two independent single-strand DNA breaks (SSBs) cause DSBs to form [62]. Computational simulation also showed consistent results where the amount of generated DSBs increased quadratically compared to the number of SSBs generated (assuming DSBs arose when two random SSBs formed within ten bases of each other) [58].

(right) are shown. DSBs are pronounced in decondensed nuclei as the irradiation dose is escalated, while DSB signals for condensed and recondensed nuclei are faint. Bar, 10 μm. (B) Quantified DSB signals obtained in A. The quadratic curve $I = kD^2$ closely fit our plotted normalized data, where I = DNA breakage caused by irradiation, D = the irradiation dose and k = the constant. (C) Bar graph showing the incidence of DNA damage when condensed, decondensed and recondensed nuclei were treated with 500 Gy of irradiation. *All data were reproduced with permission from H. Takata, T. Hanafusa, T. Mori, M. Shimura, Y. Iida, K. Ishikawa, K. Yoshikawa, Y. Yoshikawa, K. Maeshima, Chromatin compaction protects genomic DNA from radiation damage, PLoS One 8 (2013) e75622.*

We have discussed the effects on chromatin when a high dose of γ-ray irradiation is employed (Fig. 3). But what happens to chromatin when a lower dose range of irradiation is used (below 100 Gy)? To detect small amounts of DSBs, we utilized an electron-multiplying charge-coupled device (EMCCD) camera, which detects single photons. The damage suppression effects were 5-fold stronger in compacted chromatin, even at 5 Gy of irradiation when damage was observed with the EMCCD [58]. Unlike high-dose irradiation, the frequency of damage increased linearly with low dose irradiation, indicating a single step formation of DSBs that consistently aligns with the notion that increasing doses of irradiation linearly cause DSBs to form [62,63].

2.3 Chromatin compaction, but not associated protein levels, protects DNA from damage

It should be emphasized that we can directly demonstrate the protective effect of chromatin compaction using isolated nuclei in our *in vitro* system, which is relatively difficult *in vivo*. Previous work implied that chromatin-associated proteins (non-histone proteins) could protect against DNA damage [64–66], where higher protein levels had a stronger ability to prevent chromatin damage caused by ionizing irradiation. However, we found that associated proteins did not protect chromatin from damage as decondensed and recondensed chromatin possessed similar total amounts and types of associated chromatin proteins. Our finding demonstrates that the protective effect on condensed chromatin is derived from chromatin compaction itself, but not from the level of chromatin-associated proteins. We also note that our results are particularly striking compared with previous studies that showed 3.1–4.5-fold less DSBs in compacted chromatin compared to relaxed nuclear chromatin [64–66]. This is because our solid-phase system had reduced background signals (non-specific DNA breakages) and allowed higher decondensation of chromatin compared to the system using agarose-embedded cells.

2.4 Radical scavengers suppress DNA damage caused by ionizing irradiation

DNA damage often results when water molecules undergo radiolysis, creating reactive radicals [63,66–69]. The exclusion of water molecules from compacted chromatin thus generates fewer reactive radicals around

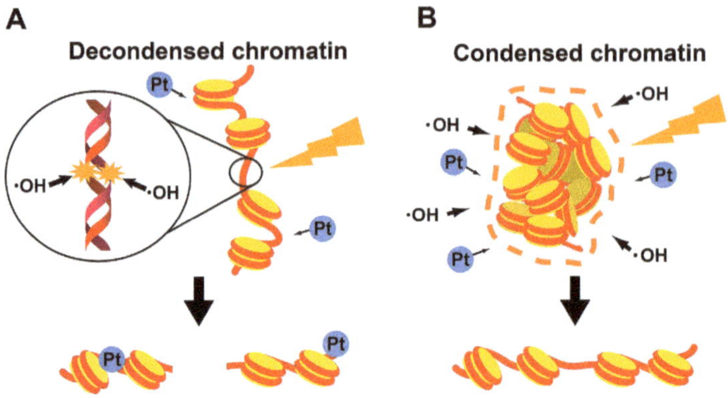

Fig. 4 Model of chromatin compaction and its suppression of DNA damage. (A) Decondensed chromatin is highly susceptible to reactive radicals generated by the radiolysis of water molecules when it is irradiated. Chemicals, like cisplatin (marked with "Pt"), are also able to better access chromatin when it is decondensed. (B) Fewer hydroxyl radicals can attack chromatin when it is more densely packed, and drugs like cisplatin, have reduced access to the chromatin. *Diagrams were reproduced and modified with permission from the authors H. Takata, T. Hanafusa, T. Mori, M. Shimura, Y. Iida, K. Ishikawa, K. Yoshikawa, Y. Yoshikawa, K. Maeshima, Chromatin compaction protects genomic DNA from radiation damage, PLoS One 8 (2013) e75622.*

the chromatin (i.e., less DNA damage) as suggested previously (Fig. 4A and B; [45,49,63,70]). Consistently, dithiothreitol (DTT; a radical scavenger) suppressed irradiation-induced DNA damage in a dose-dependent manner regardless of the condensation state of DNA [58].

2.5 Chromatin compaction protects genomic DNA against heavy ion irradiation and chemicals

Clinical applications of carbon ion radiotherapy for cancer treatments are growing worldwide (refer to chapter "Carbon ion radiation and clustered DNA double-strand breaks" by Shibata, in this issue) [71]. Therefore, we investigated if chromatin compaction could protect against DNA damage caused by heavy ion (carbon ion beam) irradiation. Compact chromatin suppressed DNA damage (7-fold) compared to decondensed chromatin. Tightly packed chromatin was able to protect DNA when exposed to either γ-rays or heavy ions [58].

Furthermore, we determined if chemical attacks could be attenuated by the compaction status of chromatin. DNA cleavage reagents, which often contain metal ions, did not apply to our study because we used ethylenediaminetetraacetic acid (EDTA) for chromatin decondensation.

Thus, we examined the effect of cisplatin, a widely used anti-tumor drug, on compacted chromatin [72]. Apoptotic cell death is induced after cisplatin generates genomic DNA covalent adducts and obstructs DNA replication and/or transcription. After treating chromatin (condensed and decondensed) with cisplatin, DNA was extracted and cisplatin adducts were quantitated using inductively coupled plasma mass spectrometry (ICP-MS). We found that the DNA in the decondensed chromatin had 10-times more cisplatin adducts (i.e. DNA damage) than the condensed chromatin, suggesting that chromatin compaction protects genomic DNA against chemical-induced damage (Fig. 4A and B; [58]).

2.6 Genome-wide survey of radiation damages

The development of next-generation sequencing technology has revolutionized genomics and allowed us to investigate the whole genome, including euchromatic and heterochromatic regions. Recent work using mouse mesothelioma cells also supports the finding that chromatin condensation shields DNA from damage [73]. The authors integrated genome-wide sequence-based information of chromatin accessibility (ATAC-seq; [74]) with locations of radiation-induced DSBs (BLISS; [75]). They found an inverse correlation between the nucleosome occupancy and the incidence of DSBs at base-pair resolution [73]. Interestingly, observed anti-correlation between the probability of DSB occurrence and nucleosome occupancy follows a power law, which is highly consistent with our *in vitro* results (Fig. 3B) [73]. This finding also supports the idea that the probability of DSB formation is proportional to the solvent-accessible surface of DNA, which may reflect the accessibility of reactive radicals and chemicals to chromatin. Overall, these studies correspond with our *in vitro* findings, which suggests the state of chromatin compaction dictates its ability to withstand DNA damage within the cell.

2.7 The evolutionary advantage of genome compaction

Previous *in vitro* work confirmed that polyamine stimulated DNA condensation also suppressed the induction of DSBs by γ-ray irradiation [42–45]. Genomic DNA compaction may be a universally conserved function for cells to maintain their genome integrity, ranging from viruses to animals. In fact, the ability of greater compaction to preserve genomic integrity would provide a selective advantage during the evolution of prokaryotes. This advantage may have been crucial during early evolutionary times, when developed DNA damage monitoring and repair systems did not exist. In eukaryotes, maintaining genomic DNA integrity is especially critical for germline cells, which bestow their genome to the next generation. For instance, yeast spores have 10-fold more condensed genomic DNA than somatic cells [76]. Human primary oocytes take 20–40 years to complete meiosis I [77] and their genome is constantly threatened by many sources of DNA damage. Interestingly they have compact chromosomes. On the other hand, rodents like mouse and rat have highly extended (dictyate-type) chromatin in their primary oocytes [78]. Indeed, it is known that human primary oocytes are much more resistant to ionizing radiation than mouse or rat primary oocytes [78], which supports our hypothesis.

3. Regulation of chromatin organization to cope with DNA damage

So far, we have discussed the importance of compact chromatin domains as a physical barrier to genotoxic stressors, keeping the chromatin less accessible to reactive radicals and chemicals. However, the formation of such domains may interfere with DNA repair because recruiting DNA repair machinery to the damage site could be restricted by limited accessibility inside the domains.

How do cells cope with this difficulty? A number of studies have suggested that cells change chromatin state following DNA damage (DNA damage response; DDR) and thereby increase chromatin accessibility [79–83]. In the following subsections, we discuss how to detect changes in the chromatin state of living cells, as well as how cells regulate their chromatin state to cope with DNA damage. Here, we focus on how the DDR regulates chromatin state. Radiation should also induce a similar DDR response, with comparable subsequent effects on the state of chromatin.

3.1 Single-nucleosome imaging/tracking discerns minute changes in the state of chromatin in living cells

We have developed single-nucleosome imaging/tracking, a powerful tool used to identify changes in the chromatin state of living cells provoked by a myriad of factors (e.g., DNA damage). This technique allows comprehensive analyses on the behavior of local chromatin [23,84–90]. In this system, cells stably expressing tagged histones (often H2B) in the nucleosomes are sparsely labeled with fluorescent dyes (Fig. 5B). Oblique

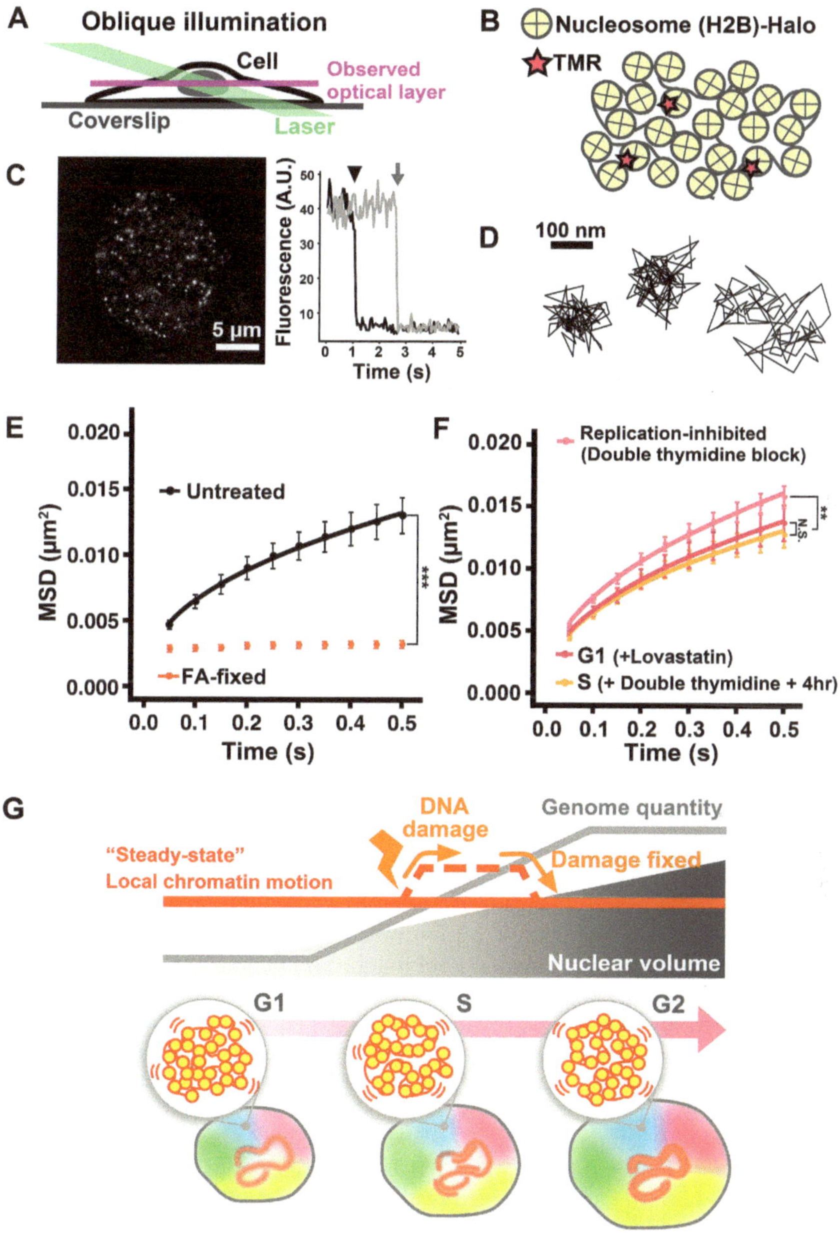

Fig. 5 Imaging of single-nucleosomes in live human cells. (A) Schematic of oblique illumination microscopy. The illumination laser (green) can excite fluorescent molecules within a limited thin optical layer (magenta) of the nucleus and reduce background noise. (B) A low concentration of HaloTag ligand TMR (red star) is used to sparsely label histone H2B tagged with HaloTag, expressed in live human cells. (C) Left: Single-nucleosome (H2B-Halo-TMR) image of a living HeLa nucleus after background

illumination microscopy is used for single-nucleosome imaging, illuminating a thin area within a single cell (Fig. 5A). The movement of distinct and well-separated dots represent individual nucleosomes (Fig. 5C) and are tracked at super-resolution for further analysis. Mean square displacement (MSD) is calculated from the nucleosome trajectory data (e.g., Fig. 5D). MSD quantitates how molecules spatially move in a specific time-period (Fig. 5E). Here, we focused on local chromatin motion for a time scale of about one second and a size scale of approximately 200 nm, corresponding to the size of chromatin domains. Single-nucleosome imaging can thus sensitively monitor local chromatin behavior, related to the compaction level of chromatin domains.

3.2 Steady-state chromatin motion and a transient rise in this motion following a DNA damage response

Recent results using single-nucleosome imaging indicated that, on average, local chromatin motion remained steady throughout G1, S and G2 phases (where the nucleus enlarges and genomic DNA doubles) in living human

subtraction. Right: Single-step photobleaching of two representative nucleosome (H2B-HaloTag-TMR) dots. The vertical axis represents the fluorescence intensity of individual TMR dots. The horizontal axis is the tracking time series. In the single-step photobleaching profile, the intensity dropped (arrowhead and arrow), suggesting that each dot represents a single H2B-HaloTag-TMR molecule in a single nucleosome. (D) Representative three trajectories of tracked single nucleosomes. (E) MSD graph ($\pm$SD among cells) comparing nucleosome motion between untreated control (black) and FA-fixed (red) HeLa cells from 0.05 to 0.5 s. (F) MSD graph ($\pm$SD among cells) comparing nucleosome motion between G1 (magenta), replication-inhibited (pale pink) and S phase (doubled thymidine block +4 h) (orange) HeLa cells from 0.05 to 0.5 s. (G) Schematic of chromatin in steady state motion throughout the cell cycle prior to mitosis. A steady state motion for local chromatin persisted throughout G1, S and G2 phases (top; red line) even while genomic quantity (i.e., doubling of genomic DNA) and nuclear volume increased. A DNA damage response initiates a transient increase in chromatin motion after perceived DNA damage (top; dashed line). This temporal change in chromatin motion presumably occurs to facilitate access for DNA repair machinery before it returns to original steady state motion levels after DNA damage is repaired or no longer detected. Five different chromosomes are highlighted with diverse colors (bottom panel). Enlarged cartoon depictions of steady-state local chromatin motion (red and yellow spheres in gray circles) are shown for G1, S and G2 phases. Note that chromatin motion appears liquid-like locally, mainly driven by thermal fluctuations. *Schematics and data were reproduced with permission from the authors S. Iida, S. Shinkai, Y. Itoh, S. Tamura, M.T. Kanemaki, S. Onami, K. Maeshima, Single-nucleosome imaging reveals steady-state motion of interphase chromatin in living human cells, Sci. Adv. 8 (2022) eabn5626.*

cells (Fig. 5G) [89]. This steady-state chromatin motion appears to be liquid-like locally, mainly driven by thermal fluctuations arising from collisions with water and other molecules in the cell [89,91]. During interphase, steady-state chromatin motion permits cells to conduct housekeeping functions (e.g., transcription and DNA replication) to occur under comparable environments [89]. Interestingly, we observed an rise in local chromatin motion in cells inhibited for DNA replication. This enhanced motion was due to the DDR, because γH2AX signals (DNA damage marker) increased in replication-inhibited cells, and ATM and ATR inhibitors were able to block the rise in local chromatin motion in response to the DDR [89].

3.3 Chromatin domain decondensation upon DNA damage response

Chromatin motion increases following a DDR, signifying decondensation of chromatin. In other words, upon DNA damage, the factor(s) downstream of ATM and/or ATR decondense chromatin and increase chromatin motion for efficient DNA repair. This result is consistent with the recent reports by dos Santos *et al.* [92], where a DDR decondensed whole chromatin and promoted rapid diffusion of proteins in the DDR.

Interestingly, the transitory rise in chromatin motion by a DDR returned to the original level 4 h following the removal of the damage inducer (Fig. 5F and G), concurrently with a diminishing of the DDR. Thus, cells can decompact chromatin domains and temporarily increase steady-state chromatin motion (~DNA accessibility) to perform their ad hoc jobs (e.g., DNA repair). The decondensed chromatin improves target searching or fosters the recruitment of pieces of DNA machinery (Fig. 5G). What occurs downstream of a DDR that causes decondensation and increased motion in chromatin? How does liquid-like chromatin behavior contribute to this process? These are intriguing questions to explore further.

4. Conclusion and clinical implications

Chromatin compaction generally acts as a physical barrier to genotoxic stressors and decreases the accessibility of reactive radicals and chemicals to the chromatin (Fig. 4A and B). It is vital to highlight that the findings discussed in this chapter provide new ways to theoretically approach combination therapies for cancers resistant to certain radio- and chemotherapies [93–96]. Moreover, because γ-rays, heavy ions and chemicals (e.g., cisplatin) more readily damage decondensed chromatin, drugs that decondense chromatin (e.g., histone deacetylase inhibitors) would complement these anti-cancer agents that act by damaging DNA.

Acknowledgments

We are grateful to K.M. Marshall for the critical reading and editing of this manuscript, S. Tamura for figure illustration, all the authors of the original work published in PLoS ONE [58], especially H. Takata and Maeshima laboratory members for helpful discussion. We must apologize that we could not mention many of the important works and articles on chromatin structure and dynamics and related methodologies due to space limitations. This work was supported by JSPS and MEXT KAKENHI grants (19H05273 and 20H05936 to K. Maeshima), the Takeda Science Foundation (to K. Maeshima), and the Uehara Memorial Foundation (to K. Maeshima). K. Minami and S. Iida are SOKENDAI Special Researchers supported by JST SPRING (JPMJSP2104 to K. Minami and S. Iida).

Competing interests

The authors declare that they have no competing interests.

References

[1] R.D. Kornberg, Chromatin structure: a repeating unit of histones and DNA, Science 184 (1974) 868–871.
[2] A.L. Olins, D.E. Olins, Spheroid chromatin units (v bodies), Science 183 (1974) 330–332.
[3] C.L. Woodcock, J.P. Safer, J.E. Stanchfield, Structural repeating units in chromatin. I. Evidence for their general occurrence, Exp. Cell Res. 97 (1976) 101–110.
[4] K. Luger, A.W. Mader, R.K. Richmond, D.F. Sargent, T.J. Richmond, Crystal structure of the nucleosome core particle at 2.8 A resolution, Nature 389 (1997) 251–260.
[5] K. Maeshima, R. Imai, S. Tamura, T. Nozaki, Chromatin as dynamic 10-nm fibers, Chromosoma 123 (2014) 225–237.
[6] C.A. Davey, D.F. Sargent, K. Luger, A.W. Maeder, T.J. Richmond, Solvent mediated interactions in the structure of the nucleosome core particle at 1.9 a resolution, J. Mol. Biol. 319 (2002) 1097–1113.
[7] K. Maeshima, S. Iida, S. Tamura, Physical nature of chromatin in the nucleus, Cold Spring Harb. Perspect. Biol. 13 (2021) a040675.
[8] J. Lipfert, S. Doniach, R. Das, D. Herschlag, Understanding nucleic acid-ion interactions, Annu. Rev. Biochem. 83 (2014) 813–841.
[9] J. Dekker, E. Heard, Structural and functional diversity of topologically associating domains, FEBS Lett. 589 (2015) 2877–2884.
[10] J.R. Dixon, S. Selvaraj, F. Yue, A. Kim, Y. Li, Y. Shen, M. Hu, J.S. Liu, B. Ren, Topological domains in mammalian genomes identified by analysis of chromatin interactions, Nature 485 (2012) 376–380.
[11] E.P. Nora, B.R. Lajoie, E.G. Schulz, L. Giorgetti, I. Okamoto, N. Servant, T. Piolot, N.L. van Berkum, J. Meisig, J. Sedat, J. Gribnau, E. Barillot, N. Bluthgen, J. Dekker, E. Heard, Spatial partitioning of the regulatory landscape of the X-inactivation centre, Nature 485 (2012) 381–385.

[12] T. Sexton, E. Yaffe, E. Kenigsberg, F. Bantignies, B. Leblanc, M. Hoichman, H. Parrinello, A. Tanay, G. Cavalli, Three-dimensional folding and functional organization principles of the Drosophila genome, Cell 148 (2012) 458–472.
[13] E. Lieberman-Aiden, N.L. van Berkum, L. Williams, M. Imakaev, T. Ragoczy, A. Telling, I. Amit, B.R. Lajoie, P.J. Sabo, M.O. Dorschner, R. Sandstrom, B. Bernstein, M.A. Bender, M. Groudine, A. Gnirke, J. Stamatoyannopoulos, L.A. Mirny, E.S. Lander, J. Dekker, Comprehensive mapping of long-range interactions reveals folding principles of the human genome, Science 326 (2009) 289–293.
[14] T. Cremer, C. Cremer, Chromosome territories, nuclear architecture and gene regulation in mammalian cells, Nat. Rev. Genet. 2 (2001) 292–301.
[15] K. Maeshima, R. Rogge, S. Tamura, Y. Joti, T. Hikima, H. Szerlong, C. Krause, J. Herman, E. Seidel, J. DeLuca, T. Ishikawa, J.C. Hansen, Nucleosomal arrays self-assemble into supramolecular globular structures lacking 30-nm fibers, EMBO J. 35 (2016) 1115–1132.
[16] B.A. Gibson, L.K. Doolittle, M.W.G. Schneider, L.E. Jensen, N. Gamarra, L. Henry, D.W. Gerlich, S. Redding, M.K. Rosen, Organization of chromatin by intrinsic and regulated phase separation, Cell 179 (2019) 470–484 e421.
[17] S. Sanulli, M.J. Trnka, V. Dharmarajan, R.W. Tibble, B.D. Pascal, A.L. Burlingame, P.R. Griffin, J.D. Gross, G.J. Narlikar, HP1 reshapes nucleosome core to promote phase separation of heterochromatin, Nature 575 (2019) 390–394.
[18] H. Strickfaden, T.O. Tolsma, A. Sharma, D.A. Underhill, J.C. Hansen, M.J. Hendzel, Condensed chromatin behaves like a solid on the mesoscale in vitro and in living cells, Cell 183 (2020) 1772–1784. e1713.
[19] K. Maeshima, T. Matsuda, Y. Shindo, H. Imamura, S. Tamura, R. Imai, S. Kawakami, R. Nagashima, T. Soga, H. Noji, K. Oka, T. Nagai, A transient rise in free Mg(2+) ions released from ATP-Mg hydrolysis contributes to mitotic chromosome condensation, Curr. Biol. 28 (2018) 444–451. e446.
[20] S.S. Rao, M.H. Huntley, N.C. Durand, E.K. Stamenova, I.D. Bochkov, J.T. Robinson, A.L. Sanborn, I. Machol, A.D. Omer, E.S. Lander, E.L. Aiden, A 3D map of the human genome at kilobase resolution reveals principles of chromatin looping, Cell 159 (2014) 1665–1680.
[21] J.C. Hansen, K. Maeshima, M.J. Hendzel, The solid and liquid states of chromatin, Epigenetics Chromatin 14 (2021) 50.
[22] Y. Hu, I. Kireev, M. Plutz, N. Ashourian, A.S. Belmont, Large-scale chromatin structure of inducible genes: transcription on a condensed, linear template, J. Cell Biol. 185 (2009) 87–100.
[23] T. Nozaki, R. Imai, M. Tanbo, R. Nagashima, S. Tamura, T. Tani, Y. Joti, M. Tomita, K. Hibino, M.T. Kanemaki, K.S. Wendt, Y. Okada, T. Nagai, K. Maeshima, Dynamic organization of chromatin domains revealed by super-resolution live-cell imaging, Mol. Cell 67 (2017) 282–293 e287.
[24] D.E. Olins, A.L. Olins, Epichromatin and chromomeres: a 'fuzzy' perspective, Open Biol. 8 (2018) 180058.
[25] J. Xu, H. Ma, J. Jin, S. Uttam, R. Fu, Y. Huang, Y. Liu, Super-resolution imaging of higher-order chromatin structures at different epigenomic states in single mammalian cells, Cell Rep. 24 (2018) 873–882.
[26] T.H. Hsieh, A. Weiner, B. Lajoie, J. Dekker, N. Friedman, O.J. Rando, Mapping nucleosome resolution chromosome folding in yeast by micro-C, Cell 162 (2015) 108–119.
[27] C. Chen, H.H. Lim, J. Shi, S. Tamura, K. Maeshima, U. Surana, L. Gan, Budding yeast chromatin is dispersed in a crowded nucleoplasm in vivo, Mol. Biol. Cell 27 (2016) 3357–3368.
[28] M. Ohno, T. Ando, D.G. Priest, V. Kumar, Y. Yoshida, Y. Taniguchi, Sub-nucleosomal genome structure reveals distinct nucleosome folding motifs, Cell 176 (2019) 520–534 e525.

[29] D.P. Hoffman, G. Shtengel, C.S. Xu, K.R. Campbell, M. Freeman, L. Wang, D.E. Milkie, H.A. Pasolli, N. Iyer, J.A. Bogovic, D.R. Stabley, A. Shirinifard, S. Pang, D. Peale, K. Schaefer, W. Pomp, C.L. Chang, J. Lippincott-Schwartz, T. Kirchhausen, D.J. Solecki, E. Betzig, H.F. Hess, Correlative three-dimensional super-resolution and block-face electron microscopy of whole vitreously frozen cells, Science 367 (2020) eaaz5357.

[30] E. Miron, R. Oldenkamp, J.M. Brown, D.M.S. Pinto, C.S. Xu, A.R. Faria, H.A. Shaban, J.D.P. Rhodes, C. Innocent, S. de Ornellas, H.F. Hess, V. Buckle, L. Schermelleh, Chromatin arranges in chains of mesoscale domains with nanoscale functional topography independent of cohesin, Sci. Adv. 6 (2020) eaba8811.

[31] K. Maeshima, K. Kaizu, S. Tamura, T. Nozaki, T. Kokubo, K. Takahashi, The physical size of transcription factors is key to transcriptional regulation in the chromatin domains, J. Phys. Condens. Matter 27 (2015) 064116.

[32] M.J. Hendzel, F. Boisvert, D.P. Bazett-Jones, Direct visualization of a protein nuclear architecture, Mol. Biol. Cell 10 (1999) 2051–2062.

[33] L. Mirny, J. Dekker, Mechanisms of chromosome folding and nuclear organization: their interplay and open questions, Cold Spring Harb. Perspect. Biol. 14 (2022) a040147.

[34] B. Bintu, L.J. Mateo, J.H. Su, N.A. Sinnott-Armstrong, M. Parker, S. Kinrot, K. Yamaya, A.N. Boettiger, X. Zhuang, Super-resolution chromatin tracing reveals domains and cooperative interactions in single cells, Science 362 (2018) eaau1783.

[35] J. Zuin, J.R. Dixon, M.I. van der Reijden, Z. Ye, P. Kolovos, R.W. Brouwer, M.P. van de Corput, H.J. van de Werken, T.A. Knoch, I.W.F. van IJcken, F.G. Grosveld, B. Ren, K.S. Wendt, Cohesin and CTCF differentially affect chromatin architecture and gene expression in human cells, Proc. Natl. Acad. Sci. U. S. A. 111 (2014) 996–1001.

[36] S.S.P. Rao, S.C. Huang, B.G.S. Hilaire, J.M. Engreitz, E.M. Perez, K.R. Kieffer-Kwon, A.L. Sanborn, S.E. Johnstone, G.D. Bascom, I.D. Bochkov, X. Huang, M.S. Shamim, J. Shin, D. Turner, Z. Ye, A.D. Omer, J.T. Robinson, T. Schlick, B.E. Bernstein, R. Casellas, E.S. Lander, E.L. Aiden, Cohesin loss eliminates all loop domains, Cell 171 (2017) 305–320. e324.

[37] G. Wutz, C. Varnai, K. Nagasaka, D.A. Cisneros, R.R. Stocsits, W. Tang, S. Schoenfelder, G. Jessberger, M. Muhar, M.J. Hossain, N. Walther, B. Koch, M. Kueblbeck, J. Ellenberg, J. Zuber, P. Fraser, J.M. Peters, Topologically associating domains and chromatin loops depend on cohesin and are regulated by CTCF, WAPL, and PDS5 proteins, EMBO J. 36 (2017) 3573–3599.

[38] T. Misteli, The self-organizing genome: principles of genome architecture and function, Cell 183 (2020) 28–45.

[39] K.F. Toth, T.A. Knoch, M. Wachsmuth, M. Frank-Stohr, M. Stohr, C.P. Bacher, G. Muller, K. Rippe, Trichostatin A-induced histone acetylation causes decondensation of interphase chromatin, J. Cell Sci. 117 (2004) 4277–4287.

[40] Q. Bian, A.S. Belmont, Revisiting higher-order and large-scale chromatin organization, Curr. Opin. Cell Biol. 24 (2012) 359–366.

[41] K.R. Kieffer-Kwon, K. Nimura, S.S.P. Rao, J. Xu, S. Jung, A. Pekowska, M. Dose, E. Stevens, E. Mathe, P. Dong, S.C. Huang, M.A. Ricci, L. Baranello, Y. Zheng, F. Tomassoni Ardori, W. Resch, D. Stavreva, S. Nelson, M. McAndrew, A. Casellas, E. Finn, C. Gregory, B.G. St Hilaire, S.M. Johnson, W. Dubois, M.P. Cosma, E. Batchelor, D. Levens, R.D. Phair, T. Misteli, L. Tessarollo, G. Hager, M. Lakadamyali, Z. Liu, M. Floer, H. Shroff, E.L. Aiden, R. Casellas, Myc regulates chromatin decompaction and nuclear architecture during B cell activation, Mol. Cell 67 (2017) 566–578 e510.

[42] M. Spotheim-Maurizot, S. Ruiz, R. Sabattier, M. Charlier, Radioprotection of DNA by polyamines, Int. J. Radiat. Biol. 68 (1995) 571–577.
[43] R.L. Warters, G.L. Newton, P.L. Olive, R.C. Fahey, Radioprotection of human cell nuclear DNA by polyamines: radiosensitivity of chromatin is influenced by tightly bound spermine, Radiat. Res. 151 (1999) 354–362.
[44] T. Douki, Y. Bretonniere, J. Cadet, Protection against radiation-induced degradation of DNA bases by polyamines, Radiat. Res. 153 (2000) 29–35.
[45] Y. Yoshikawa, T. Mori, N. Magome, K. Hibino, K. Yoshikawa, DNA compaction plays a key role in radioprotection against double-strand breaks as revealed by single-molecule observation, Chem. Phys. Lett. 456 (2008) 80–83.
[46] M. Suzuki, C. Crozatier, K. Yoshikawa, T. Mori, Y. Yoshikawa, Protamine-induced DNA compaction but not aggregation shows effective radioprotection against double-strand breaks, Chem. Phys. Lett. 480 (2009) 113–117.
[47] R. Scully, A. Panday, R. Elango, N.A. Willis, DNA double-strand break repair-pathway choice in somatic mammalian cells, Nat. Rev. Mol. Cell Biol. 20 (2019) 698–714.
[48] S.E. Polo, S.P. Jackson, Dynamics of DNA damage response proteins at DNA breaks: a focus on protein modifications, Genes Dev. 25 (2011) 409–433.
[49] M. Falk, E. Lukasova, S. Kozubek, Chromatin structure influences the sensitivity of DNA to gamma-radiation, Biochim. Biophys. Acta 1783 (2008) 2398–2414.
[50] I.G. Cowell, N.J. Sunter, P.B. Singh, C.A. Austin, B.W. Durkacz, M.J. Tilby, gammaH2AX foci form preferentially in euchromatin after ionising-radiation, PLoS One 2 (2007) e1057.
[51] T.C. Karagiannis, K.N. Harikrishnan, A. El-Osta, Disparity of histone deacetylase inhibition on repair of radiation-induced DNA damage on euchromatin and constitutive heterochromatin compartments, Oncogene 26 (2007) 3963–3971.
[52] J.A. Kim, M. Kruhlak, F. Dotiwala, A. Nussenzweig, J.E. Haber, Heterochromatin is refractory to gamma-H2AX modification in yeast and mammals, J. Cell Biol. 178 (2007) 209–218.
[53] R.S. Vasireddy, T.C. Karagiannis, A. El-Osta, gamma-radiation-induced gammaH2AX formation occurs preferentially in actively transcribing euchromatic loci, Cell. Mol. Life Sci. 67 (2010) 291–294.
[54] I. Revet, L. Feeney, S. Bruguera, W. Wilson, T.K. Dong, D.H. Oh, D. Dankort, J.E. Cleaver, Functional relevance of the histone gammaH2Ax in the response to DNA damaging agents, Proc. Natl. Acad. Sci. U. S. A. 108 (2011) 8663–8667.
[55] T. Kurashige, M. Shimamura, Y. Nagayama, Differences in quantification of DNA double-strand breaks assessed by 53BP1/gammaH2AX focus formation assays and the comet assay in mammalian cells treated with irradiation and N-acetyl-L-cysteine, J. Radiat. Res. 57 (2016) 312–317.
[56] A. Cole, Chromosome structure: theoretical, Biophysics 1 (1967) 305–375.
[57] Y. Gavrieli, Y. Sherman, S.A. Ben-Sasson, Identification of programmed cell death in situ via specific labeling of nuclear DNA fragmentation, J. Cell Biol. 119 (1992) 493–501.
[58] H. Takata, T. Hanafusa, T. Mori, M. Shimura, Y. Iida, K. Ishikawa, K. Yoshikawa, Y. Yoshikawa, K. Maeshima, Chromatin compaction protects genomic DNA from radiation damage, PLoS One 8 (2013) e75622.
[59] J.C. Hansen, Conformational dynamics of the chromatin fiber in solution: determinants, mechanisms, and functions, Annu. Rev. Biophys. Biomol. Struct. 31 (2002) 361–392.
[60] K. Maeshima, M. Eltsov, Packaging the genome: the structure of mitotic chromosomes, J. Biochem. 143 (2008) 145–153.

[61] K. Maeshima, U.K. Laemmli, A two-step scaffolding model for mitotic chromosome assembly, Dev. Cell 4 (2003) 467–480.
[62] W.J. Cannan, D.S. Pederson, Mechanisms and consequences of double-strand DNA break formation in chromatin, J. Cell. Physiol. 231 (2016) 3–14.
[63] A. Valota, F. Ballarini, W. Friedland, P. Jacob, A. Ottolenghi, H.G. Paretzke, Modelling study on the protective role of OH radical scavengers and DNA higher-order structures in induction of single- and double-strand break by gamma-radiation, Int. J. Radiat. Biol. 79 (2003) 643–653.
[64] R.L. Warters, B.W. Lyons, Variation in radiation-induced formation of DNA double-strand breaks as a function of chromatin structure, Radiat. Res. 130 (1992) 309–318.
[65] M.C. Elia, M.O. Bradley, Influence of chromatin structure on the induction of DNA double strand breaks by ionizing radiation, Cancer Res. 52 (1992) 1580–1586.
[66] J. Nygren, M. Ljungman, G. Ahnstrom, Chromatin structure and radiation-induced DNA strand breaks in human cells: soluble scavengers and DNA-bound proteins offer a better protection against single- than double-strand breaks, Int. J. Radiat. Biol. 68 (1995) 11–18.
[67] R.E. Krisch, M.B. Flick, C.N. Trumbore, Radiation chemical mechanisms of single- and double-strand break formation in irradiated SV40 DNA, Radiat. Res. 126 (1991) 251–259.
[68] J.R. Milligan, J.A. Aguilera, J.F. Ward, Variation of single-strand break yield with scavenger concentration for plasmid DNA irradiated in aqueous solution, Radiat. Res. 133 (1993) 151–157.
[69] Y. Yoshikawa, K. Hizume, Y. Oda, K. Takeyasu, S. Araki, K. Yoshikawa, Protective effect of vitamin C against double-strand breaks in reconstituted chromatin visualized by single-molecule observation, Biophys. J. 90 (2006) 993–999.
[70] J.F. Ward, J.R. Milligan, R.C. Fahey, Factors Controlling the Radiosensitivity of Cellular DNA, Royal Society of Chemistry, 1997. Place Published.
[71] H. Tsujii, T. Kamada, A review of update clinical results of carbon ion radiotherapy, Jpn. J. Clin. Oncol. 42 (2012) 670–685.
[72] D. Wang, S.J. Lippard, Cellular processing of platinum anticancer drugs, Nat. Rev. Drug Discov. 4 (2005) 307–320.
[73] F. Brambilla, J.M. Garcia-Manteiga, E. Monteleone, L. Hoelzen, A. Zocchi, A. Agresti, M.E. Bianchi, Nucleosomes effectively shield DNA from radiation damage in living cells, Nucleic Acids Res. 48 (2020) 8993–9006.
[74] J.D. Buenrostro, B. Wu, H.Y. Chang, W.J. Greenleaf, ATAC-seq: a method for assaying chromatin accessibility genome-wide, Curr. Protoc. Mol. Biol. 109 (2015) 21.29.21–21.29.29.
[75] W.X. Yan, R. Mirzazadeh, S. Garnerone, D. Scott, M.W. Schneider, T. Kallas, J. Custodio, E. Wernersson, Y. Li, L. Gao, Y. Federova, B. Zetsche, F. Zhang, M. Bienko, N. Crosetto, BLISS is a versatile and quantitative method for genome-wide profiling of DNA double-strand breaks, Nat. Commun. 8 (2017) 15058.
[76] J.M. Bryant, J. Govin, L. Zhang, G. Donahue, B.F. Pugh, S.L. Berger, The linker histone plays a dual role during gametogenesis in Saccharomyces cerevisiae, Mol. Cell. Biol. 32 (2012) 2771–2783.
[77] T. Sadler, Langman's Medical Embryology, eleventh ed., Lippincott Williams & Wilkins, 2010.
[78] T.G. Baker, P. Neal, Action of ionizing radiations on the mammalian ovary, in: S. Zuckerman, B.J. Weir (Eds.), The Ovary, Academic Press, New York and London, 1977, pp. 1–58.
[79] M.J. Kruhlak, A. Celeste, G. Dellaire, O. Fernandez-Capetillo, W.G. Muller, J.G. McNally, D.P. Bazett-Jones, A. Nussenzweig, Changes in chromatin structure and mobility in living cells at sites of DNA double-strand breaks, J. Cell Biol. 172 (2006) 823–834.

[80] M. Murga, I. Jaco, Y. Fan, R. Soria, B. Martinez-Pastor, M. Cuadrado, S.M. Yang, M.A. Blasco, A.I. Skoultchi, O. Fernandez-Capetillo, Global chromatin compaction limits the strength of the DNA damage response, J. Cell Biol. 178 (2007) 1101–1108.
[81] A. Seeber, M.H. Hauer, S.M. Gasser, Chromosome dynamics in response to DNA damage, Annu. Rev. Genet. 52 (2018) 295–319.
[82] H. Lu, J. Saha, P.J. Beckmann, E.A. Hendrickson, A.J. Davis, DNA-PKcs promotes chromatin decondensation to facilitate initiation of the DNA damage response, Nucleic Acids Res. 47 (2019) 9467–9479.
[83] V.O. Chagin, B. Reinhart, A. Becker, O. Mortusewicz, K.L. Jost, A. Rapp, H. Leonhardt, M.C. Cardoso, Processive DNA synthesis is associated with localized decompaction of constitutive heterochromatin at the sites of DNA replication and repair, Nucleus 10 (2019) 231–253.
[84] S. Hihara, C.G. Pack, K. Kaizu, T. Tani, T. Hanafusa, T. Nozaki, S. Takemoto, T. Yoshimi, H. Yokota, N. Imamoto, Y. Sako, M. Kinjo, K. Takahashi, T. Nagai, K. Maeshima, Local nucleosome dynamics facilitate chromatin accessibility in living mammalian cells, Cell Rep. 2 (2012) 1645–1656.
[85] R. Nagashima, K. Hibino, S.S. Ashwin, M. Babokhov, S. Fujishiro, R. Imai, T. Nozaki, S. Tamura, T. Tani, H. Kimura, M. Shribak, M.T. Kanemaki, M. Sasai, K. Maeshima, Single nucleosome imaging reveals loose genome chromatin networks via active RNA polymerase II, J. Cell Biol. 218 (2019) 1511–1530.
[86] J. Lerner, P.A. Gomez-Garcia, R.L. McCarthy, Z. Liu, M. Lakadamyali, K.S. Zaret, Two-parameter mobility assessments discriminate diverse regulatory factor behaviors in chromatin, Mol. Cell 79 (2020) 677–688 e676.
[87] P.A. Gomez-Garcia, S. Portillo-Ledesma, M.V. Neguembor, M. Pesaresi, W. Oweis, T. Rohrlich, S. Wieser, E. Meshorer, T. Schlick, M.P. Cosma, M. Lakadamyali, Mesoscale modeling and single-nucleosome tracking reveal remodeling of clutch folding and dynamics in stem cell differentiation, Cell Rep. 34 (2021) 108614.
[88] Y. Itoh, S. Iida, S. Tamura, R. Nagashima, K. Shiraki, T. Goto, K. Hibino, S. Ide, K. Maeshima, 1,6-Hexanediol rapidly immobilizes and condenses chromatin in living human cells, Life Sci. Alliance 4 (2021) e202001005.
[89] S. Iida, S. Shinkai, Y. Itoh, S. Tamura, M.T. Kanemaki, S. Onami, K. Maeshima, Single-nucleosome imaging reveals steady-state motion of interphase chromatin in living human cells, Sci. Adv. 8 (2022) eabn5626.
[90] S. Ide, S. Tamura, K. Maeshima, Chromatin behavior in living cells: lessons from single-nucleosome imaging and tracking, BioEssays 44 (2022) e2200043.
[91] S.S. Ashwin, et al., Organization of fast and slow chromatin revealed by single-nucleosome dynamics, Proc. Natl. Acad. Sci. U. S. A. 116 (2019) 19939–19944, https://doi.org/10.1073/pnas.1907342116.
[92] A. Dos Santos, A.W. Cook, R.E. Gough, M. Schilling, N.A. Olszok, I. Brown, L. Wang, J. Aaron, M.L. Martin-Fernandez, F. Rehfeldt, C.P. Toseland, DNA damage alters nuclear mechanics through chromatin reorganization, Nucleic Acids Res. 49 (2021) 340–353.
[93] M.S. Kim, M. Blake, J.H. Baek, G. Kohlhagen, Y. Pommier, F. Carrier, Inhibition of histone deacetylase increases cytotoxicity to anticancer drugs targeting DNA, Cancer Res. 63 (2003) 7291–7300.
[94] A. Abbas, S. Gupta, The role of histone deacetylases in prostate cancer, Epigenetics 3 (2008) 300–309.
[95] F. Zhang, T. Zhang, Z.H. Teng, R. Zhang, J.B. Wang, Q.B. Mei, Sensitization to gamma-irradiation-induced cell cycle arrest and apoptosis by the histone deacetylase inhibitor trichostatin A in non-small cell lung cancer (NSCLC) cells, Cancer Biol. Ther. 8 (2009) 823–831.
[96] H. De Schutter, S. Nuyts, Radiosensitizing potential of epigenetic anticancer drugs, Anti Cancer Agents Med. Chem. 9 (2009) 99–108.

CHAPTER FOUR

Mechanical force induced DNA double-strand breaks: Ultrasound

Yue Ma[a,b,*] and Iwaki Akiyama[a]

[a]Faculty of Life and Medical Science, Doshisha University, Kyotanabe, Japan
[b]Institute of Biomedical Engineering, Chinese Academy of Medical Sciences and Peking Union Medical College, Tianjin, China
[*]Corresponding author: e-mail address: mayue@bme.cams.cn

Contents

Abstract

Since the application of ultrasound for clinical diagnosis and therapeutic purposes has been increased rapidly, the effects of exposure to ultrasound on DNA molecules were studied. In this chapter, we introduced various effects of DNA damages caused by different conditions of exposure of ultrasound. Ultrasound with different sound pressure and pulse transmission conditions have been applied in our study. We discussed the threshold of sound pressure of ultrasound-induced DNA damages. Different kinds of pulses of ultrasound and microbubbles' influences on DNA double-strand breaks were also shown.

1. Introduction

With the development of medical ultrasonic technology, application of ultrasound for clinical diagnosis and therapeutic purposes has been increased rapidly. The characteristics of ultrasound waves are used to determine the morphological features of organs and blood flow velocity for

The Enzymes, Volume 51
ISSN 1874-6047
https://doi.org/10.1016/bs.enz.2022.08.004

diagnosis [1–6]. Especially, from the outbreak of COVID-19, ultrasonic equipment is becoming even more important for diagnosing the progression of pneumonia, since it can be used bedside [5]. For therapeutic purposes, various ultrasound techniques have been developed to promote wound healing, transdermal drug delivery, targeted drug delivery [6–8], etc. Depending on the aim and requirements, different types of ultrasonic waves have been applied [9–13]. Diagnostic applications of ultrasound usually use short pulses (~μs) [9–12]. In contrast, continuous or long-pulse ultrasonic waves are more common in therapeutic applications [9,10,12,13].

DNA damage is usually categorized into four types: base changes, cross-linking, and single- and double-strand breaks (DSBs). DSBs are considered to be the most serious since they can potentially lead to cancer and cell death [14]. Currently, several methodologies can be used to detect DSBs both in vivo and in vitro [14–21], such as polymerase chain reaction, immunological assays, in situ hybridization, etc. Despite the availability of these methods, it is difficult to evaluate the number of double-strand breaks in a quantitative manner, especially for long genome-sized DNA molecules [22–24]. Recently, it has been demonstrated that the direct visualization of single giant DNA molecules using fluorescence microscopy can provide useful information on the structure and function of genomic DNA molecules in solution [25–28], and can be used to analyze DSBs in a quantitative manner. The DNA utilized in present study is T4 GT7 bacteriophage DNA (166 kbp). Compared to plasmid DNA, it is long enough to show the similar semiflexible properties of human eukaryotic DNA (ca. 105–106 kbp).

Since the medical applications of ultrasound are quickly expanding and DNA is one of the most important biomolecules in living organisms, our group had investigated the effects of exposure to ultrasound on DNA molecules [28–31]. In this chapter, we will introduce different effects of DNA damage caused by different kinds of ultrasound.

2. Different ultrasound induced DNA damages

2.1 Threshold of sound pressure for DNA double-strand breaks

After 1.0 MHz ultrasound exposure, the double-strand breaks in giant DNA molecules were tested at the level of single DNA molecules. DNA samples were exposed to various types of ultrasonic pulsed waves with different pulse repetition periods and pulse durations [30]. As shown in Fig. 1, The results regarding DNA damage induced by 1.0 MHz ultrasound exposure also

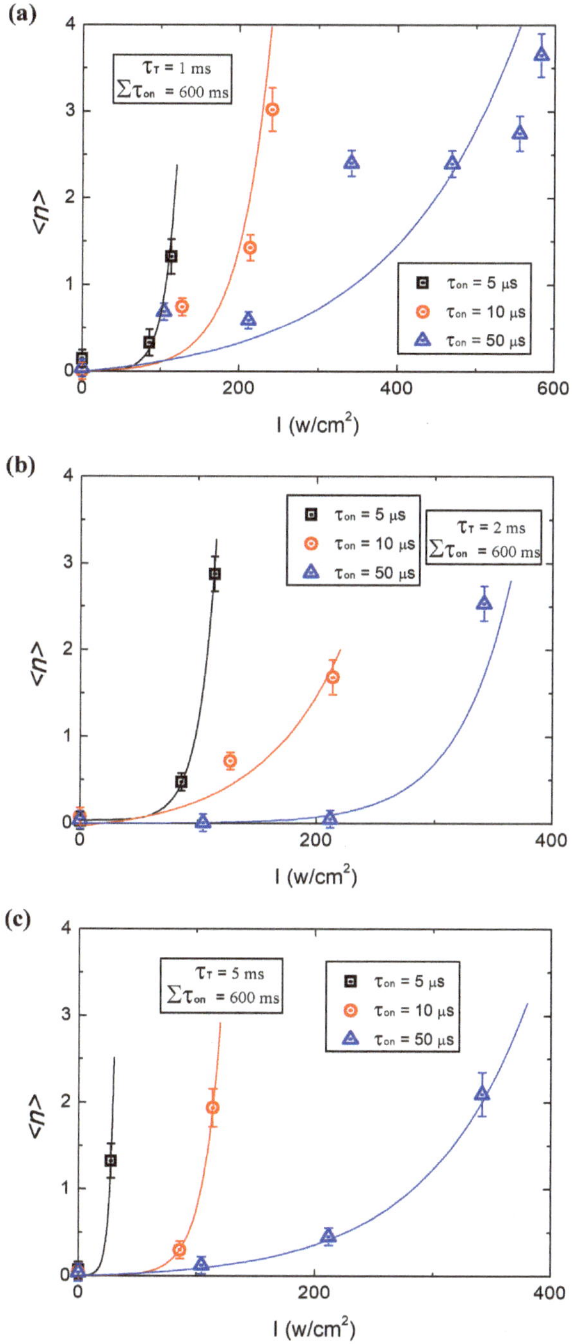

Fig. 1 Number of double-strand breaks per DNA molecule, $<n>$, vs. the intensity of ultrasound, I, under the same total pulse duration, $\sum \tau_{on} = 600$ ms. The pulse repetition period of ultrasound was (A) 1 ms, (B) 2 ms, or (C) 5 ms. *Y. Ma, K. Ishihara, K. Yoshida, I. Akiyama, K. Yoshikawa, Double-strand breaks in genome-sized DNA caused by megahertz ultrasound, J. Acoust. Soc. Am. 150 (1) (2021) 241–247.*

establish the existence of a threshold value of sound pressure for DNA double-strand breaks. Below this threshold, the probability of double-strand breaks is essentially zero. Cavitations generated by ultrasound was thought to be the main reason which caused DNA double-strand breaks. Based on the experimental evidence on the existence of threshold value of sound pressure, it is expected that DNA damage was also induced by cavitations.

Similar experiment was also performed to evaluate the protection of ascorbic acid (AA) on ultrasound-induced DNA double-strand breaks (DSBs) [31]. Fig. 2A exemplifies the DNA observations after exposure to ultrasound. Based on the data on the sound pressure level-dependence of the average length $<L>$ given in Fig. 2B, we obtained a graph of the average

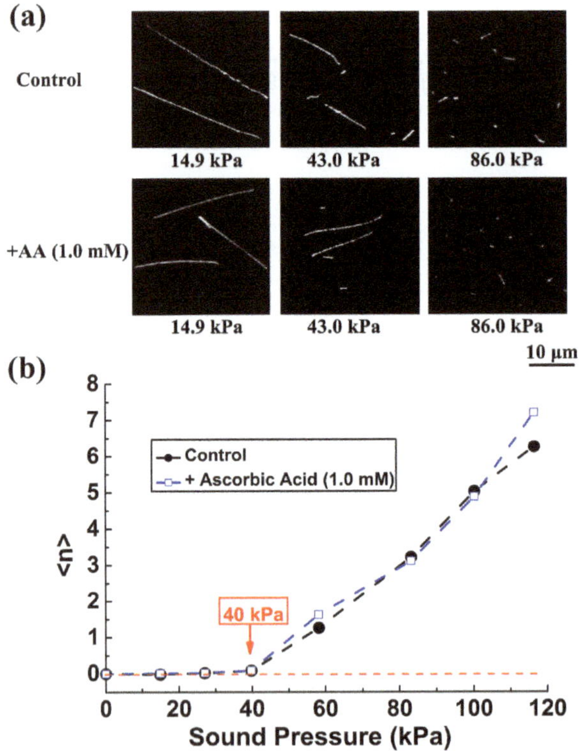

Fig. 2 Ultrasound-induced DSBs. (A) Fluorescence microscopic images of DNA molecules fixed on a glass surface after exposure to ultrasound at different sound pressures. (B) Average lengths of DNA, $<L>$, *vs.* the sound pressure of ultrasound. *Y. Ma, N. Ogawa, Y. Yoshikawa, T. Mori, T. Imanaka, Y. Watanabe, K. Yoshikawa, Protective effect of ascorbic acid against double-strand breaks in giant DNA: marked differences among the damage induced by photo-irradiation, gamma-rays and ultrasound, Chem. Phys. Lett. 638 (2015): 205–209.*

number of breaks per original DNA molecule, $<n>$, with respect to the sound pressure, as shown in Fig. 2C. This confirms the existence of a threshold value of sound pressure for DNA double-strand breaks; below this threshold, the probability of double-strand breaks is essentially zero. Above the threshold, $<n>$ increases linearly with the sound pressure. The nature of the increase in $<n>$ is almost the same independent of the presence or absence of AA. These results demonstrate that AA does not have an obvious protective effect against ultrasound-induced DSBs.

2.2 The influence of pulse of ultrasound on DSBs

We obtained plots of the average number of breaks per original DNA molecule with respect to the sound pressure, as shown in Fig. 1 [30]. For ultrasound with the same pulse repetition period, the threshold sound pressure decreased as the pulse duration increased. Beyond the threshold value of sound pressure, damage to DNA increased as the sound pressure increased. In terms of the intensity of ultrasound, the double-strand breaks of DNA also increased as the intensity increased. However, the risk of DNA damage decreased as the pulse duration of ultrasound increased.

With regard to ultrasound with the same pulse duration, the threshold of sound pressure increased as the pulse repetition period of ultrasound increased. As for the intensity of ultrasound, the risk of DNA damage decreased as the frequency of ultrasound was increased.

The above results demonstrated that, after exposure to ultrasound with the same total pulse duration, the threshold value of sound pressure decreased as the pulse duration increased, and increased as the pulse repetition period increased. The risk of DNA damage was also higher for ultrasound with a shorter pulse duration and longer pulse repetition period.

2.3 Microbubbles enhanced DSBs induced by ultrasound

To evaluate the effects of microbubbles on DNA double-strand breaks caused by ultrasound exposure, we observed DNA samples with different concentrations of microbubbles exposed to ultrasound with a pulse repetition period of 1 ms, a pulse duration of 10 μs and a total pulse duration, $\sum \tau_{on} = 600$ ms. For microbubbles, we used Sonazoid, which is a commercially available ultrasound contrast agent. The sound pressure level-dependence of the DNA double-strand breaks is given in Fig. 3 [30].

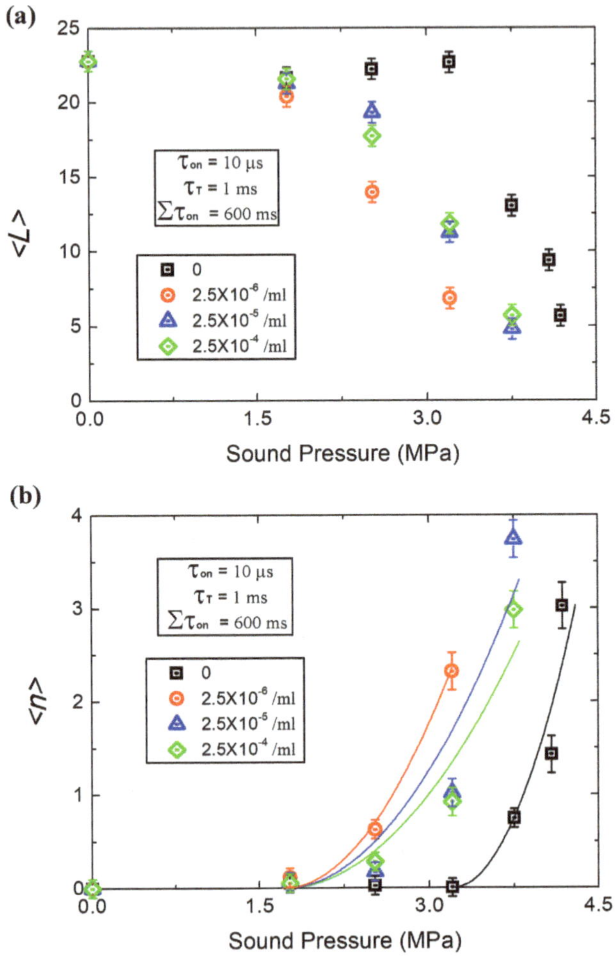

Fig. 3 Effects of microbubbles on ultrasound-induced DSBs. (A) Average lengths of DNA, $<L>$, vs. the sound pressure of ultrasound; (B) Number of DNA double-strand breaks per molecule, $<n>$, vs. the sound pressure of ultrasound. *Y. Ma, K. Ishihara, K. Yoshida, I. Akiyama, K. Yoshikawa, Double-strand breaks in genome-sized DNA caused by megahertz ultrasound, J. Acoust. Soc. Am. 150 (1) (2021) 241–247.*

After microbubbles were added to the DNA samples, in terms of the sound pressure of ultrasound, the DSBs of DNA were more serious with a higher concentration of microbubbles. Although the threshold sound pressure decreased by ca. 50% after microbubbles were added to the samples, the concentration of microbubbles does not appear to strongly influence the threshold.

3. Experimental system for ultrasound exposure

3.1 Measurement of sound pressure

In this study, it is necessary to measure the sound pressure by the exposure to ultrasonic waves in advance. In particular, since DNA is infused into a silicon tube and exposure to ultrasonic waves, the sound pressure must be measured inside the silicon tube. The sound pressure was measured by a needle type hydrophone (HNR0500, ONDA, Sunnyvale, CA) which is equipped with a PZT disc shaped transducer (diameter: 0.5 mm). Inner and outer diameter of the Silicone tube are 2.0 mm and 3.0 mm, respectively. Diameter and focal length of the transducer are 40 mm and 50 mm, respectively. A hole was made through the tube for navigating to the center of inner tube as shown in Fig. 4. Waves received by hydrophone were observed by an oscilloscope (DL850E, Yokogawa, Japan). The voltage signals applied to the transducer were burst sinusoidal waves which were generated and amplified by a signal generator (Keysight, 33,210, Santa Rosa, CA) and a power amplifier (A150, Electronics & Innovation, Rochester, NY), respectively. Frequency of sinusoidal waves was 1.0 MHz. Burst cycles were 10.

Negative amplitude of sound pressures in degassed pure water without tube were measured in the range of 1.6–4.0 MPa. Conductivity of the pure water is less than 10 μS/cm and dissolved oxygen concentration of the degassed water is less than 1.0 mg/L. After the measurement without tube, sound pressures in the tube were also measured. The measured values of negative sound pressure amplitude are plotted in the range of applied voltage from 100 to 250 V_{pp} as shown in Fig. 5. It is shown that the difference between measured values in tube and those without tube is less than 10% as shown in Fig. 5. Because the uncertainty of hydrophone calibration is 10% according to the manufacturer's report, it could conclude that the fluctuation of sound pressure is less than 10%.

3.2 Mapping of sound pressure distribution

Two-dimensional distribution of sound pressure formed by transmitting ultrasonic waves from the concave focused transducer was measured by a hydrophone. The hydrophone (HNP0200, ONDA, Sunnyvale, Ca) is a needle type and is equipped with a PVDF transducer disc (diameter: 0.2 mm). A step motorized xyz-axes stage attached with the hydrophone

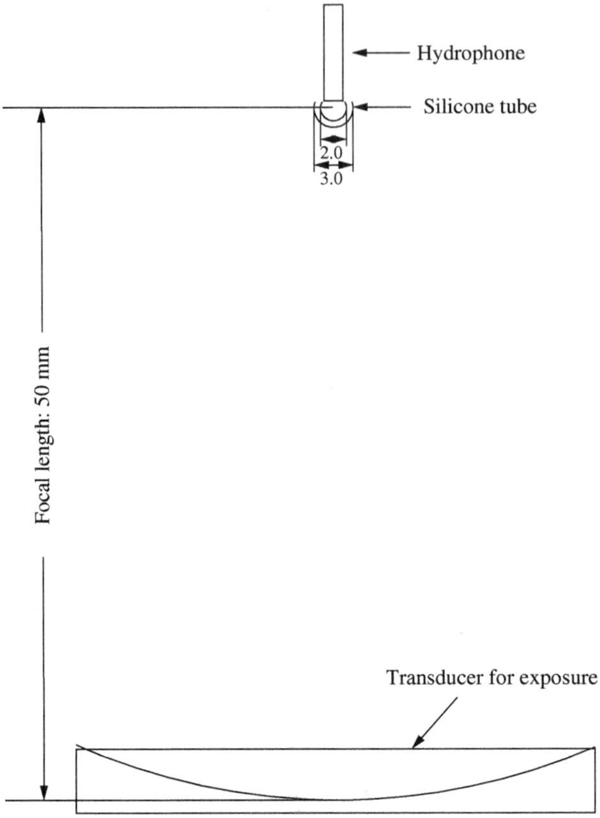

Fig. 4 Sound pressure in the silicone tube was measured by a hydrophone (HNR0500, Onda, USA). The transducer for exposure was concave type. The focal length is 50 mm and diameter of the transducer is 40 mm.

was controlled to move in x–z plane by a step of 0.2 mm in x-axis and 0.5 mm in z-axis, which covered the area of 8.0 mm × 30.0 mm. The x-axis was set along the radial direction of the transducer and the z-axis was set along the ultrasound propagation direction. A water tank was filled with the degassed pure water and the sound pressure at each voltage applied to the transducer was measured by the hydrophone and the reception system as described in the previous section.

The measured map of sound pressure distribution is shown in Fig. 6 [30]. As a result, it is confirmed that the beam width at focal region is 2.0 mm in x-axis as shown in Fig. 6.

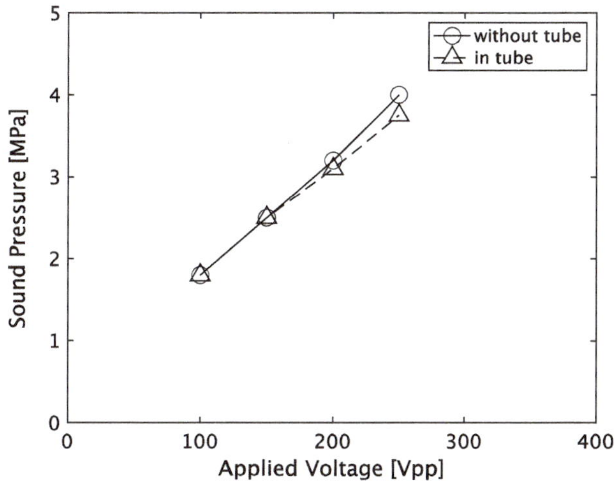

Fig. 5 Plots of sound pressure to applied voltage to the transducer. Sound pressure was measured by a hydrophone in the silicone tube and without the tube.

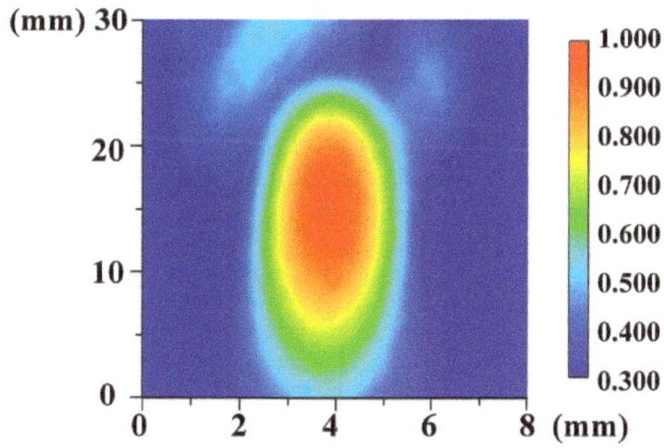

Fig. 6 Two-dimensional map of sound pressure distribution obtained by scanning a hydrophone (HNP0200, Onda, Sunnyvale CA) in x-z plane passing through the focal point. *Y. Ma, K. Ishihara, K. Yoshida, I. Akiyama, K. Yoshikawa, Double-strand breaks in genome-sized DNA caused by megahertz ultrasound, J. Acoust. Soc. Am. 150 (1) (2021) 241–247.*

4. Summary

DNA damage induced by ultrasound exposure establish the existence of a threshold value of sound pressure for DNA double-strand breaks. Below this threshold, the probability of double-strand breaks is essentially zero. Cavitation generated by ultrasound was thought to be the main reason which caused DNA double-strand breaks.

For the same total pulse duration of ultrasound, the risk of DNA damage was higher for ultrasound with a shorter pulse duration and longer pulse repetition period. The threshold value of sound pressure decreased as the pulse duration increased, and increased as the pulse repetition period increased.

The DSBs of DNA were more serious with a higher concentration of microbubbles. Although the threshold sound pressure decreased by ca. 50% after microbubbles were added to the samples, the concentration of microbubbles does not appear to strongly influence the threshold.

Acknowledgment

This work was supported in part by the MEXT-Supported Program for the Strategic Research Foundation at Private Universities, 2013-2017 and by JSPS KAKENHI Grant Number JP20H01877. The authors thanks Mr. Kazuya Ishihara and Dr. Kenji Yoshida for their experimental contribution, and Dr. Kenichi Yoshikawa for his crucial discussion.

References

[1] D.L. Miller, N.B. Smith, M.R. Bailey, G.J. Czarnota, K. Hynynen, I.R.S. Makin, et al., Overview of therapeutic ultrasound applications and safety considerations, J. Ultrasound Med. 31 (4) (2012) 623–634.

[2] M.A. Buldakov, M.A. Hassan, P. Jawaid, N.V. Cherdyntseva, T. Kondo, Cellular effects of low-intensity pulsed ultrasound and X-irradiation in combination in two human leukaemia cell lines, Ultrason. Sonochem. 23 (2015) 339–346.

[3] M.P. Krafft, Overcoming inactivation of the lung surfactant by serum proteins: a potential role for fluorocarbons? Soft Matter 11 (30) (2015) 5982–5994.

[4] G.R. Harris, C.C. Church, D. Dalecki, M.C. Ziskin, J.E. Bagley, Comparison of thermal safety practice guidelines for diagnostic ultrasound exposures, Ultrasound Med. Biol. 42 (2) (2016) 345–357.

[5] J.S. Abramowicz, J. Basseal, World Federation for Ultrasound in Medicine and Biology Position Statement: how to perform a safe ultrasound examination and clean equipment in the context of COVID-19, Ultrasound Med. Biol. 46 (7) (2020) 1821–1826.

[6] R.J. Siegal, S. Vaezy, R. Martin, L. Crum, Therapeutic ultrasound, Part II*. High intensity focused ultrasound: a method of hemostasis, Echocardiography 18 (4) (2001) 309–315.

[7] S. Mitragotri, Healing sound: the use of ultrasound in drug delivery and other therapeutic applications, Nat. Rev. Drug Discov. 4 (3) (2005) 255–260.

[8] K. Milowska, T. Gabryelak, Enhancement of ultrasonically induced cell damage by phthalocyanines in vitro, Ultrasonics 48 (8) (2008) 724–730.

[9] K.G. Baker, V.J. Robertson, F.A. Duck, A review of therapeutic ultrasound: biophysical effects, Phys. Ther. 81 (7) (2001) 1351–1358.

[10] S. Behrens, K. Spengos, M. Daffertshofer, H. Schroeck, C.E. Dempfle, M. Hennerici, Transcranial ultrasound-improved thrombolysis: diagnostic vs. therapeutic ultrasound, Ultrasound Med. Biol. 27 (12) (2001) 1683–1689.

[11] D. Dalecki, Mechanical bioeffects of ultrasound, Annu. Rev. Biomed. Eng. 6 (2004) 229–248.

[12] P. Dijkmans, L. Juffermans, R. Musters, A. van Wamel, F. Ten Cate, W. van Gilst, et al., Microbubbles and ultrasound: from diagnosis to therapy, Eur. J. Echocardiogr. 5 (4) (2004) 245–246.

[13] G. Ter Haar, Therapeutic applications of ultrasound, Prog. Biophys. Mol. Biol. 93 (1–3) (2007) 111–129.
[14] J.H. Hoeijmakers, DNA damage, aging, and cancer, New Eng. J. Med. 361 (15) (2009) 1475–1485.
[15] J. Bradbury, S. Jackson, The Complex Matter of DNA Double-Strand Break Detection, Portland Press Ltd., London, 2003.
[16] M. Fenech, Chromosomal damage rate, aging, and diet a, Ann. N. Y. Acad. Sci. 854 (1) (1998) 23–36.
[17] S. Loft, H.E. Poulsen, Cancer risk and oxidative DNA damage in man, J. Mol. Med. 74 (6) (1996) 297–312.
[18] T. Rich, R.L. Allen, A.H. Wyllie, Defying death after DNA damage, Nature 407 (6805) (2000) 777–783.
[19] T. Weinert, DNA damage and checkpoint pathways: molecular anatomy and interactions with repair, Cell 94 (5) (1998) 555–558.
[20] C. Fant, A. Granzotto, J.-L. Mestas, J. Ngo, M. Lafond, C. Lafon, et al., DNA double-strand breaks in murine mammary tumor cells induced by combined treatment with doxorubicin and controlled stable cavitation, Ultrasound Med. Biol. 47 (10) (2021) 2941–2957.
[21] Y. Furusawa, T. Kondo, K. Tachibana, L.B. Feril Jr., Ultrasound-induced DNA damage and cellular response: historical review, mechanisms analysis, and therapeutic implications, Radiat. Res. 197 (6) (2022) 662–672.
[22] S. Kumari, R.P. Rastogi, K.L. Singh, S.P. Singh, R.P. Sinha, DNA damage: detection strategies, EXCLI J. 7 (2008) 44–62.
[23] W. Liao, M.A. McNutt, W.-G. Zhu, The comet assay: a sensitive method for detecting DNA damage in individual cells, Methods 48 (1) (2009) 46–53.
[24] G.I. McFadden, In Situ Hybridization. Methods in Cell Biology, Elsevier, Amsterdam, 1995, pp. 165–183.
[25] S.F. Shimobayashi, T. Iwaki, T. Mori, K. Yoshikawa, Probability of double-strand breaks in genome-sized DNA by γ-ray decreases markedly as the DNA concentration increases, J. Chem. Phys. 138 (17) (2013). 05B604_1.
[26] Y. Yoshikawa, T. Mori, M. Suzuki, T. Imanaka, K. Yoshikawa, Comparative study of kinetics on DNA double-strand break induced by photo-and gamma-irradiation: protective effect of water-soluble flavonoids, Chem. Phys. Lett. 501 (1–3) (2010) 146–151.
[27] Y. Yoshikawa, M. Suzuki, N. Yamada, K. Yoshikawa, Double-strand break of giant DNA: protection by glucosyl-hesperidin as evidenced through direct observation on individual DNA molecules, FEBS Lett. 566 (1–3) (2004) 39–42.
[28] K. Yoshida, N. Ogawa, Y. Kagawa, H. Tabata, Y. Watanabe, T. Kenmotsu, et al., Effect of low-frequency ultrasound on double-strand breaks in giant DNA molecules, Appl. Phys. Lett. 103 (6) (2013), 063705.
[29] R. Kubota, Y. Yamashita, T. Kenmotsu, Y. Yoshikawa, K. Yoshida, Y. Watanabe, et al., Double-strand breaks in genome-sized DNA caused by ultrasound, ChemPhysChem 18 (8) (2017) 959–964.
[30] Y. Ma, K. Ishihara, K. Yoshida, I. Akiyama, K. Yoshikawa, Double-strand breaks in genome-sized DNA caused by megahertz ultrasound, J. Acoust. Soc. Am. 150 (1) (2021) 241–247.
[31] Y. Ma, N. Ogawa, Y. Yoshikawa, T. Mori, T. Imanaka, Y. Watanabe, et al., Protective effect of ascorbic acid against double-strand breaks in giant DNA: marked differences among the damage induced by photo-irradiation, gamma-rays and ultrasound, Chem. Phys. Lett. 638 (2015) 205–209.

CHAPTER FIVE

DNA damage and biological responses induced by Boron Neutron Capture Therapy (BNCT)

Natsuko Kondo*
Particle Radiation Oncology Research Center, Institute for Integrated Radiation and Nuclear Science, Kyoto University, Osaka, Japan
*Corresponding author: e-mail address: kondo.natsuko.3x@kyoto-u.ac.jp

Contents

Abstract

Boron Neutron Capture Therapy (BNCT) is a tumor cell selective high LET (linear energy transfer) particle beam therapy. The patient is administrated a boron (^{10}B) compound via intravenous injection or infusion, and when ^{10}B is sufficiently accumulated in the tumor, neutron beams containing epithermal neutrons as the main component are irradiated. Epithermal neutrons lose energy in the body and become thermal neutrons. The captured ^{10}B undergoes a (n, α) reaction with thermal neutrons, and the resulting α particles and ^{7}Li nuclei have short ranges of 9–10μm and 4–5μm, respectively, and do not reach the surrounding cells in normal tissues. Therefore, these high LET—heavy charged particles can selectively kill cancer cells. The cell-killing effect of these heavy charged particles is thought to be triggered by DNA damage. It is known that DNA damage caused by heavy charged particles is more serious and difficult to

The Enzymes, Volume 51
ISSN 1874-6047
https://doi.org/10.1016/bs.enz.2022.08.005

repair than DNA damage caused by Low LET radiation such as X-rays and γ-rays. This review focuses on DNA damage, e.g., DNA strand breaks and DNA damage repair caused by BNCT and describes the resulting biological response.

1. Introduction

1.1 Mechanism of BNCT

Boron Neutron Capture Therapy (BNCT) is a tumor selective particle radiation therapy using high- linear energy transfer (LET) radiations produced by nuclear capture and fission reactions. BNCT consists of the binary approach. First, the stable isotope boron-10 (^{10}B)-containing drug is administered to the patient to obtain a sufficient tumor ^{10}B concentration. For successful treatment, ~20–35 μg/g of ^{10}B must be selectively delivered to the tumor cells. Second, the patient is irradiated with enough neutrons (~10^{12} n/cm^2) mainly consisting of epithermal neutrons from an external radiation source [1]. The epithermal neutrons are moderated to thermal neutrons in the body. The resulting $^{10}B(n,\alpha)^7Li$ capture reaction with the thermal neutrons produces alpha particle and lithium-7 nucleus. The ranges of these high LET particles are short (9–10 μm and 4–5 μm, for α particles and ^7Li nuclei, respectively), so tumor cells are selectively killed while sparing adjacent normal tissues as shown in Fig. 1. Unlike other particle beam therapies such as proton beam and carbon beam, which use the Bragg peak to physically concentrate the dose on the cancer tissue, BNCT is a particle beam therapy that uses the biological feature of selective boron drug accumulation in the tumor to deliver the dose. Therefore, there are high expectations for BNCT against infiltrative or disseminated tumors into normal tissues such as malignant gliomas which are difficult to control with other particle therapy methods.

1.2 Neutron beam component

Neutron beams are mixed beams which contain fast neutrons, thermal neutrons, low-LET gamma-rays. Therefore, besides the dose from capture reactions in ^{10}B: [$^{10}B(n,\alpha)^7Li$], the background dose consists of (a) low LET gamma rays resulting from the capture of thermal neutrons by hydrogen atoms in tissue: [$^1H(n,\gamma)^2H$] but also present as incident gammas in the beam, (b) high-LET protons, either recoiling hydrogen nuclei due to collisions with fast neutrons: [$^1H(n,n')^1H$] or ejected due to capture of low-energy neutrons in nitrogen: [$^{14}N(n,p)^{14}C$], and (c) high-LET heavier charged

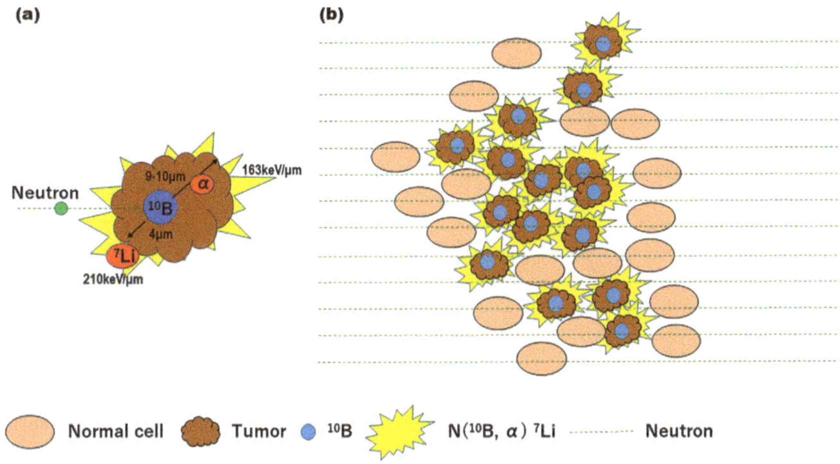

Fig. 1 Schema of mechanistic action of BNCT (A) $^{10}B(n,\alpha)^{7}$ Li capture reaction occurring in a tumor cell taken up by ^{10}B. When ^{10}B is irradiated by low energy thermal neutron, the resulting $^{10}B(n,\alpha)^{7}$ Li capture reaction produces high linear energy transfer (LET) α particles (^{4}He) and recoiling lithium-7 (^{7} Li) atoms with short path lengths 9–10, and 4 μm within the diameter of a single cell, respectively (B) The $^{10}B(n,\alpha)^{7}$ Li capture reactions mainly occur in the tumor cells which include ones invading into normal tissues. Thus, BNCT can selectively kill the tumor cells and spare the surrounding normal cells.

particles ^{14}C, released as products of capture reactions in ^{14}N [$^{14}N(n,p)^{14}C$] [2]. The background dose is much lower compared with the dose from capture reactions in ^{10}B: [$^{10}B(n,\alpha)^{7}Li$] [2]. However, as long as normal tissue receives the background dose, BNCT protocols should ideally be maximize the ratio between the tumor selective boron dose and the non-selective background dose [1,2].

1.3 ^{10}Boron compounds in clinical use

Ideal boron compounds would (a) be able to maintain the enough boron concentration in tumor tissues during neutron irradiation, (b) have systemic low or non-toxicity and be cleared rapidly from normal tissues and blood after neutron irradiation, and (c) not accumulate in normal tissues and guarantee the high ratio of "concentration in tumor tissue/concentration in normal tissue." Various low and high molecular weight boron compounds have been developed and their efficacy has been shown *in vitro* and in experimental animal tumor models, however only two boron compounds have been mainly used in clinical BNCT studies [1]. One is sodium mercaptoundecahydro-*closo*-dodecaborate ($Na_2B_{12}H_{11}SH$), commonly known as sodium borocaptate or BSH and the other is, a boron-containing amino acid

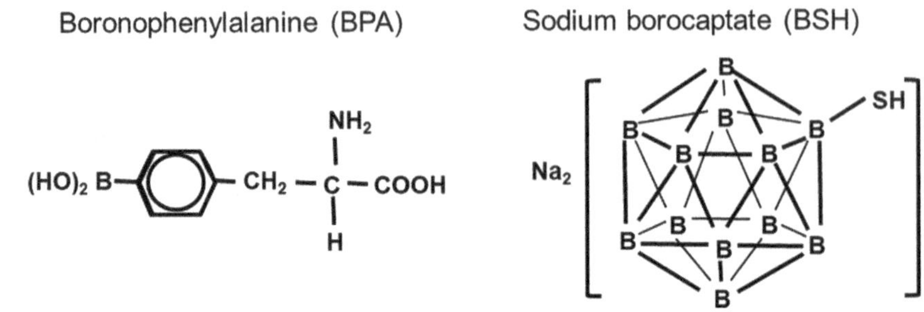

Fig. 2 L-4-dihydroxy-borylphenylalanine, BPA (boronophenylalanine) and undecahydro-mercapto-*closo*-dodecaborate, BSH (sodium borocaptate).

(L)-4-dihydroxy-borylphenylalanine, known as boronophenylalanine or BPA (Fig. 2). BSH was used clinically for malignant glioma by Hatanaka and Nakagawa in Japan for the first time and showed significant antitumor effects [3,4]. Successively, phase I/II clinical trial for malignant glioma was performed in Europe [5]. Although BSH is not selectively taken up by tumor cells, it is considered to diffusely accumulate in brain tumor stroma where the blood-brain barrier is disrupted. On the other hand, BPA is selectively taken up by tumor cell taken up via L-Type Amino Acid Transporter 1 (LAT1), and LAT1 expression has been reported to be higher in various cancer types than in cells of normal tissue [6]. BPA was clinically introduced for melanoma BNCT by Mishima et al [7] and was demonstrated to be taken up by other histologic type of tumor, rat brain tumor, 9L gliosarcoma by Coderre et al. [8]. Increased water solubility as BPA-fructose complex (BPA–F) [9] promoted its clinical use for malignant glioma in the United States [10,11], Finland [12], Sweden [13], and Japan [14] and it became clear that BPA was more therapeutically effective than BSH. After that, BPA was evaluated successfully in BNCT for head and neck cancers [15,16] and then became the central Boron compound in BNCT.

1.4 Neutron irradiation field and neutron source: Transition from reactor to accelerator-based systems

Since the first BNCT of gliomas at BNL in 1951, BNCT has used thermal neutrons from nuclear reactors in United States or Japan. In 1994, a modification of the reactor irradiation field at BNL made it possible to use epithermal neutrons [17]. Since the late 1990s, BNCT using epithermal neutrons has become possible in nuclear reactors and enabled BNCT clinical studies in Europe [18] and Japan [19] of not only brain tumor but also head and neck cancers [15] or other organ cancer such as mesotheliomas in lung [20], sarcomas in limbs [21,22]. Basic and clinical research was also conducted in Argentina, China, Taiwan, and Korea using epithermal neutrons from reactors [23]. In parallel with the successful BNCT clinical studies in nuclear reactors, the development of accelerator neutron sources has progressed around the world. In 2008, the Cyclotron Irradiation System for BNCT (C-BENS) was installed at KURNS, Japan in collaborative work with Sumitomo Heavy Industries, Ltd [24]. Clinical trials using this system for malignant glioma and recurrent head and neck cancer were conducted for the first time in the world, from October 2012 to April 2018, with favorable results [25,26]. In 2020, this accelerator system was approved as a medical device. At the same time, BPA was also approved as borofalan (^{10}B), a boron drug for BNCT in 2020. In Japan, the National Cancer Center is currently conducting clinical trials for malignant melanoma and angiosarcoma using accelerator BNCT system with a solid lithium target-linac-vertical beam [27]. In Finland, Korea, and China, clinical trials with accelerator based BNCT are under preparation [28].

In order to develop BNCT into a safer and more effective therapy, which will be in increasing demand in the future with the availability of accelerator-based therapies, it is very important to know the biological effects of BNCT resulting from DNA damage, and following DNA repair, checkpoint, and cell death.

2. DNA damage and repair mechanism induced by BNCT

Radiation damages various parts of the cell, including the cell membrane, cytoplasm, and nucleus, but DNA damage is thought to be the most important cause of mitotic death. Low and high linear energy transfer ionizing radiations induce a plethora of DNA damage, and the proportion of the locally clustered damage increases with LET of the radiation [29]. Isolated DNA lesions (mainly induced by low LET radiation), including DSBs, SSB, and damaged bases located at a distance from other damage, are generally repaired efficiently. High-LET radiation induces complex DNA damage, a unique class of DNA lesions that includes two or more individual lesions within one or two helical turns of the DNA. Complex DNA lesions are more difficult to repair than isolated lesions and in some instances are irreparable [30].

The cell-killing effect of BNCT is supposed to result from DNA damage caused by high linear energy transfer (LET) radiation, i.e. alpha particles and

Li recoil nuclei produced by the capture reaction of a thermal neutron by a ^{10}B atom, [$^{10}B(n,\alpha)^7Li$]. The DNA damages induced by BNCT were investigated using plasmid DNA and boric acid [31,32]. These studies indicated that SSBs and DSBs were produced and increased as the absorbed dose increased in the BNCT mixed field. The DSB/SSB ratio for the boric acid consisting of pure ^{10}B isotopes with 0.3 M was 3–4 times greater than that for boric acids with low concentration, 0.03 M[32]. High-LET alpha and ^7Li particles were considered to be produced in larger amounts by the boron neutron capture reaction [^{10}B (n, a) ^7Li] and consequently the DSB/SSB ratio was increased for the boric acids with high concentrations [32].

One of the earliest responses to an ionizing radiation-induced DSB is the phosphorylation of a histone, H2AX, at serine 139, yielding a focal product (gamma-H2AX) that can be detected by a fluorescent antibody. Other proteins, such as TP53BP1, then colocalize at these sites of DSBs and so become candidate signatures for processing and repair of this damage. These nuclear foci are strikingly apparent after their induction by ionizing radiation and can be counted [33]. Comparing with gamma ray, the numbers of gamma-H2AX, or TP53BP1 produced by BNCT were not different at 30 min after irradiation, while they became higher after 2–24 h after irradiation [34,35]. And, the analysis of focus size showed that foci are larger in cells treated by BNCT 30 or 120 min after irradiation as compared to those in the gamma-ray or neutron only groups [34,35]. These results indicated boron neutron capture reaction [^{10}B (n, a) ^7Li] induced complex DSBs [34,35].

In mouse intracranial glioblastoma model, the gamma H2AX foci in brain tumor disappeared in saline treated mouse, i.e., in the absence of ^{10}B, 24 h after neutron irradiation. However, in the tumor of 500 mg/kg BPA-treated mouse, i.e., in the presence of ^{10}B, the gamma H2AX foci remained (4 foci/cell), indicating that unrepairable DSBs were induced by [$^{10}B(n,\alpha)^7Li$] [36] (Fig. 3A and B). In this model, the gamma H2AX foci produced by the component of contaminated gamma-ray dose were investigated and found that they disappeared 24 h after irradiation (Fig. 4) in the tumor, indicating that only the gamma H2AX foci produced by [$^{10}B(n,\alpha)^7Li$] remained in the tumor after BPA-BNCT [36]. From these results, the [$^{10}B(n,\alpha)^7Li$] can induce severe and unrepairable DSBs, suggesting BNCT has a remarkable effect on radiation-resistant glioblastoma. Similar results were reported by rat lymphosarcoma xenograft model. In the study, 20 h after BPA-BNCT, the positivity of staining with gamma H2AX and Poly(ADP-ribose) which is another immediate marker for both single

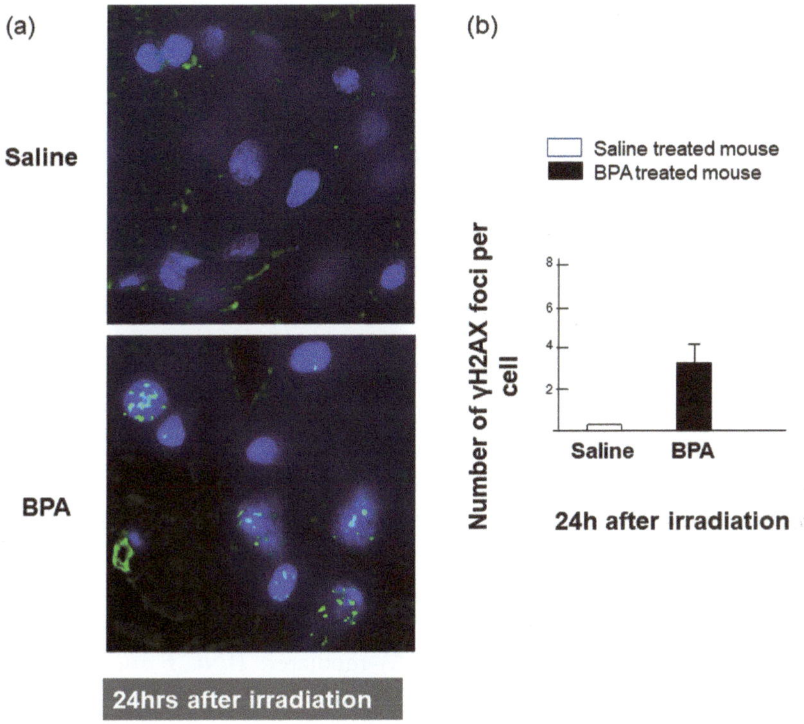

Fig. 3 γH2AX foci in brain tumor model after neutron irradiation. (A) Representative images of nuclear γH2AX foci of brain tumor cells. Upper image shows saline-treated mouse, and lower shows BPA treated mouse. DAPI, staining of nuclear DNA. (B) The number of γH2AX foci 24h after irradiation. Bars represent the standard errors. *Reproduced with permission. Copyright 2016, Elsevier N. Kondo, H. Michiue, Y. Sakurai, Detection of γH2AX foci in mouse normal brain and brain tumor after boron neutron capture therapy, Rep. Pract. Oncol. Radiother. 21 (2016) 108–112.*

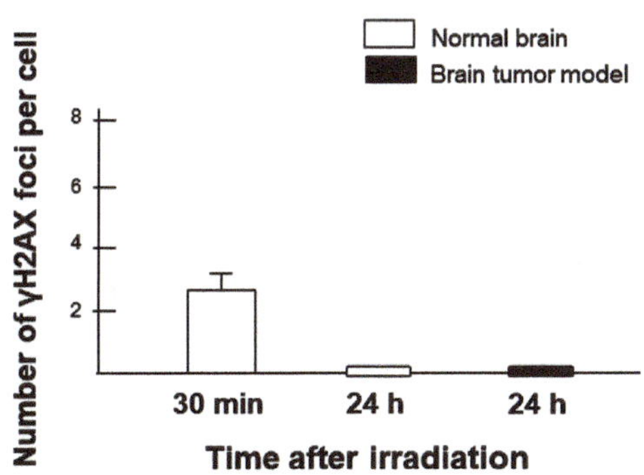

Fig. 4 γH2AX foci in normal brain and brain tumor model after γ-irradiation. The number of γH2AX foci 24h after irradiation at the times indicated post-irradiation. *Reproduced with permission. Copyright 2016, Elsevier N. Kondo, H. Michiue, Y. Sakurai, Detection of γH2AX foci in mouse normal brain and brain tumor after boron neutron capture therapy, Rep. Pract. Oncol. Radiother. 21 (2016) 108–112.*

strand break (SSB) and DSB, synthesized by poly(ADP-ribose) polymerase (PARP) in nucleus, was higher than that of saline-treated group (without ^{10}B) [37]

DNA repair mechanisms induced after BNCT were investigated by *in vitro* studies. The two major DSB repair pathways (homologous recombination repair: HRR and non-homologous end-joining: NHEJ) have been analyzed. HRR uses the sister chromatid in the late S and G2/M phase in cell cycle, therefore completes accurate repair. Error-prone NHEJ can function in all the time of cell cycle, including G0/G1 phase where most of somatic cells belong [33]. NHEJ repair protein deficient cell, Ku80 [34] and DNA ligase IV [38], showed higher sensitivity to the [^{10}B(n,α)^{7}Li], compared to the wild or parental cells, and Ku80 and DNA ligase IV were supposed to contribute to the repair of DSBs caused by BNCT. Depending on cancer cell lines, the proportion of contributory DSB repair pathways induced by BPA-BNCT may be different. In the human undifferentiated thyroid follicular cancer cell line, representing the Rad51 and Rad54, HRR was predominantly activated, while in the human melanoma cell line, both HRR and Ku70, on behalf of NHEJ were activated [35]. In the hepatocellular carcinoma, boric acid (BA)-mediated BNCT induced G2/M arrest, and HRR was activated [39]. More detailed studies are necessary regarding DSB repair pathway involved in BNCT.

3. Biological effects of BNCT

Mild DNA damage normally leads to the induction of cell-cycle arrest, whereas severe and irreparable injury shifts the cellular response towards induction of the senescence or cell death programs, such as apoptosis, mitotic catastrophe (MC), autophagy and necrosis [40]. BNCT causes severe and irreparable DNA damage as stated before, and is expected to induce these responses.

3.1 Apoptosis

P53, as one of the main targets of ATM/ATR, is a universal sensor of genotoxic stress and has a key role in cellular responses to DNA damage, determining the fate of cells towards either survival, which involves cell-cycle delay accompanied by repair of DNA damage, or cell death [40]. *P53* mutation occurs frequently in tumor cells and that makes tumor cells more resistant to radiation therapy [41]. *P53 mt* squamous cell carcinoma (SAS), and glioblastoma cells were more resistant to BNCT, comparing with *p53 wild* tumor cells [41,42]. Regarding apoptosis, in *p53 wild* SAS cells, G1 arrest and apoptosis occurred after 6 hours, while in *p53mt* SAS cells, G1 arrest did not occur, only G2/M arrest occurred after 12 h, and apoptosis occurred after 48 h [42]. Thus, BNCT causes both p53 dependent apoptosis and p53 independent apoptosis. Similarly, both *p53 wild* and *mt* glioblastoma cells were induced apoptosis by BNCT [43].

Glioma stem cells, which are also known to resistant to X-ray and chemotherapy, take up BPA [44] and BNCT induced apoptosis via mitochondrial pathway [45].

3.2 Mitotic catastrophe

To maintain genome integrity, cells respond to DNA damage by either a delay in cell-cycle progression, allowing time for proper DNA repair, or by the elimination of cells that are irreparably injured. When checkpoints are compromised, cells can enter mitosis prematurely before the completion of DNA repair and initiate MC [40]. MC is not an ultimate manifestation of cell death but rather a process leading to apoptosis or necrosis [46]. MC occurs either during or shortly after a dysregulated/failed mitosis and can be accompanied by morphological alterations, including micronucleation (which often results from chromosomes and/or chromosome fragments that have not been distributed evenly between the daughter nuclei) and multinucleation (the presence of two or more nuclei of similar or heterogeneous sizes, derived from a deficient separation during cytokinesis) [40]. Micronuclei were observed after BNCT with BPA in melanoma [47], and with BPA or BSH in EL4 lymphoma, SCC VII squamous cell carcinoma, FM3A mammary carcinoma, and EMT6/KU sarcoma, followed by apoptosis [48]. Chromosomal aberrations, which also lead to MC, increased in human lymphocytes after BSH or BPA-BNCT in a dose dependent manner [49].

Although BPA-BNCT suppressed the growth of both SAS/*neo* and SAS/*mp53* tumors to undetectable levels in vivo, half of SAS/*mp53* tumors showed regrowth and histological examination of BNCT-treated tumors revealed chromosomal condensation, micronucleation, nuclear segmentation and intra- and intercellular vacuolation [50]. Notably, multinucleated giant cells appeared in SAS/*mp53* tumors early after BNCT, suggesting mitotic catastrophe. In SAS/*mp53* tumors treated with BNCT, a rapid decrease in phosphorylated cell division cycle 2 (cdc2) and a high level of cyclin B1, required for premature mitosis, were observed [50].

3.3 Necrosis

Necrosis is an acute form of cell death usually initiated following energy loss. Until recently, necrosis was considered to be an unregulated form of cell death; however, accumulating experimental data demonstrate that, except under extreme conditions, necrosis may be a genetically regulated process. Poly(ADPribosyl)ation was shown to have a central role in modulating the cellular response to severe genotoxic stress and induction of cell death through necrotic mechanisms [40]. After DNA damage, the activation of poly(ADP)-ribose polymerase (PARP) regulates the translocation of HMGB1, the immunostimulatory histone binding protein, from the nucleus to the cytosol [51]. BPA-BNCT increased Poly(ADPribosyl) foci in xenograft tumor tissue slightly at 6h, and remarkably at 20h post irradiation, and HMGB1 expression in the tumor tissue 6 hours after irradiation [37]. In another study, at 3 days post-BPA-based BNCT irradiation in a SAS xenograft mouse model, plasma HMGB1 levels were higher than those in the non-irradiation control, and HMGB1 was detected in both nuclei and cytoplasm in tumor cells [52]. Additionally, increased plasma HMGB1 levels post-BNCT irradiation were detected even when tumors decreased in size. Therefore, BNCT induced necrosis with HMGB1 release, and HMGB1could be a biomarker for BNCT for evaluating the therapeutic response during BNCT [52].

3.4 Abscopal effect

Abscopal effect indicates 'remission of metastases outside the field of irradiation' and has occurred very rarely before the immune-checkpoint inhibitors in combination radiation therapy were introduced [53]. Abscopal effect must be mediated by an antitumor immune response. The main effectors of tumor cell killing are cytotoxic T lymphocytes, which rely on neoantigen presentation by dendritic cells [53]. Neoantigens are abundantly produced by DNA damaging agents, radiation or chemotherapy. Exposure of radiation -induced micronuclear DNA to cytoplasm is the potent trigger of type I interferon signaling and contribute to the onset of abscopal effect [53].

BNCT was firstly shown by Trivillin et al., to cause abscopal effects using colon cancer xenografts [54] and this abscopal effect was enhanced by immune-therapy agent Bacillus Calmette-Guerin (BCG) [55]. The mechanism is not fully elucidated, but BNCT-induced micronuclear DNA would trigger the abscopal effects by BNCT. Misrepaired DNA damage after BNCT might be the source of tumor neoantigens and contribute to the abscopal effects by BNCT.

4. Closing remarks

BNCT can induces apoptosis, mitotic catastrophe, and necrosis due to irreparable DNA damage in a tumor cell-specific manner by [$^{10}B(n,\alpha)^{7}Li$]. These powerful biological effects should lead to the excellent results of BNCT even against the most miserable tumor, glioblastoma, or refractory head and neck tumors. Recently, we have obtained the accelerator-based BNCT and more cancer patients will have chance to choose BNCT. In addition, further research will develop more tumor-selective boron compounds, sophisticated drug delivery or better combined therapy (for example, immuno-modulating agents) with BNCT. Clinical and basic research on BNCT will become more and more necessary and noteworthy in the future.

Reference

[1] R.F. Barth, P. Mi, W. Yang, Boron delivery agents for neutron capture therapy of cancer, Cancer Commun. (Lond.) 38 (2018) 35.
[2] J.A. Coderre, J.C. Turcotte, K.J. Riley, Boron neutron capture therapy: cellular targeting of high linear energy transfer radiation, Technol. Cancer Res. Treat 2 (2003) 355–375.
[3] A.H. Soloway, H. Hatanaka, M.A. Davis, Penetration of brain and brain tumor. VII. Tumor-binding sulfhydryl boron compounds, J. Med. Chem. 10 (1967) 714–717.
[4] H. Hatanaka, Y. Nakagawa, Clinical results of long-surviving brain tumor patients who underwent boron neutron capture therapy, Int. J. Radiat. Oncol. Biol. Phys. 28 (1994) 1061–1066.
[5] M.J. Vos, B. Turowski, F.E. Zanella, Radiologic findings in patients treated with boron neutron capture therapy for glioblastoma multiforme within EORTC trial 11961, Int. J. Radiat. Oncol. Biol. Phys. 61 (2005) 392–399.
[6] P. Häfliger, R.P. Charles, The L-type amino acid transporter LAT1-an emerging target in cancer, Int. J. Mol. Sci. 20 (2019) 2428.
[7] Y. Mishima, H. Kondoh, Dual control of melanogenesis and melanoma growth: overview molecular to clinical level and the reverse, Pigment Cell Res. 13 (2000) 10–22.
[8] J.A. Coderre, J.D. Glass, R.G. Fairchild, Selective delivery of boron by the melanin precursor analogue p-boronophenylalanine to tumors other than melanoma, Can. Res. 50 (1990) 138–141.
[9] K. Yoshino, A. Suzuki, Y. Mori, Improvement of solubility of p-boronophenylalanine by complex formation with monosaccharides, Strahlenther Onkol. 165 (1989) 127–129.
[10] M. Chadha, J. Capala, J.A. Coderre, Boron neutron-capture therapy (BNCT) for glioblastoma multiforme (GBM) using the epithermal neutron beam at the Brookhaven National Laboratory, Int. J. Radiat. Oncol. Biol. Phys. 40 (1998) 829–834.
[11] A.Z. Diaz, Assessment of the results from the phase I/II boron neutron capture therapy trials at the Brookhaven National Laboratory from a clinician's point of view, J. Neurooncol. 62 (2003) 101–109.
[12] L. Kankaanranta, T. Seppala, H. Koivunoro, L-Boronophenylalanine-mediated boron neutron capture therapy for malignant glioma progressing after external beam radiation therapy: a phase I study, Int. J. Radiat. Oncol. Biol. Phys. 80 (2011) 369–376.

[13] R. Henriksson, J. Capala, A. Michanek, Swedish Brain Tumour Study Group, Boron neutron capture therapy (BNCT) for glioblastoma multiforme: a phase II study evaluating a prolonged high-dose of boronophenylalanine (BPA), Radiother Oncol. 88 (2008) 183–191.
[14] S. Miyatake, S. Kawabata, R. Hiramatsu, Boron neutron capture therapy for malignant brain tumors, Neurol. Med. Chir. (Tokyo) 56 (2016) 361–371.
[15] L. Kankaanranta, T. Seppälä, H. Koivunoro, Boron neutron capture therapy in the treatment of locally recurred head-and-neck cancer: final analysis of a phase I/II trial, Int. J. Radiat. Oncol. Biol. Phys. 82 (2012) e67–e75.
[16] L.W. Wang, Y.W. Chen, C.Y. Ho, Fractionated boron neutron capture therapy in locally recurrent head and neck cancer: a prospective phase I/II trial, Int. J. Radiat. Oncol. Biol. Phys. 95 (2016) 396–403.
[17] D.N. Slatkin, J.M. Javid, D.D. Joel, A history of 20th-century boron neutron-capture therapy, J. Neurol. Neurobiol. 3 (2017).
[18] A. Wittig, R.L. Moss, F. Stecher-Rasmussen, Neutron activation of patients following boron neutron capture therapy of brain tumors at the high flux reactor (HFR) Petten (EORTC Trials 11961 and 11011), Strahlenther. Onkol. 181 (2005) 774–782.
[19] Y. Sakurai, H. Tanaka, T. Takata, Advances in boron neutron capture therapy (BNCT) at Kyoto University—from reactor-based BNCT to accelerator-based BNCT, J. Korean Phys. Soc. 67 (2015) 76–81.
[20] M. Suzuki, K. Endo, H. Satoh, A novel concept of treatment of diffuse or multiple pleural tumors by boron neutron capture therapy (BNCT), Radiother. Oncol. 88 (2008) 192–195.
[21] T. Fujimoto, M. Suzuki, S. Kuratsu, BNCT for primary synovial sarcoma, Appl. Radiat. Isot. 169 (2021), 109407.
[22] T. Fujimoto, M. Suzuki, T. Sudo, Boron neutron capture therapy for clear cell sarcoma, Appl. Radiat. Isot. 166 (2020), 109324.
[23] P.R. Menéndez, B.M. Roth, M.D. Pereira, BNCT for skin melanoma in extremities: updated Argentine clinical results, Appl. Radiat. Isot. 67 (2009) S50–S53.
[24] H. Tanaka, Y. Sakurai, M. Suzuki, Experimental verification of beam characteristics for cyclotron-based epithermal neutron source (C-BENS), Appl. Radiat. Isotopes 69 (2011) 1642–1645.
[25] S. Kawabata, M. Suzuki, K. Hirose, Accelerator-based BNCT for patients with recurrent glioblastoma: a multicenter phase II study, Neurooncol. Adv. 3 (2021) vdab067.
[26] K. Hirose, A. Konno, J. Hiratsuka, Boron neutron capture therapy using cyclotron-based epithermal neutron source and borofalan (^{10}B) for recurrent or locally advanced head and neck cancer (JHN002): An open-label phase II trial, Radiother. Oncol. 155 (2021) 182–187.
[27] H. Igaki, N. Murakami, S. Nakamura, Scalp angiosarcoma treated with linear accelerator-based boron neutron capture therapy: a report of two patients, Clin. Transl. Radiat. Oncol. 33 (2022) 128–133.
[28] Nuclear Technology Review, Austria (2020) 46–48.
[29] H. Nikjoo, P. O'Neill, W.E. Wilson, Computational approach for determining the spectrum of DNA damage induced by ionizing radiation, Radiat. Res. 156 (2001) 577–583.
[30] M.E. Lomax, L.K. Folkes, P. O'Neill, Biological consequences of radiation-induced DNA damage: relevance to radiotherapy, Clin. Oncol. (R. Coll. Radiol.) 25 (2013) 578–585.
[31] E. Okamoto, T. Yamamoto, K. Nakai, Detection of DNA double-strand breaks in boron neutron capture reaction, Appl. Radiat. Isot. 106 (2015) 185–188.
[32] E. Jamsranjav, A. Ito, Y. Kato, DNA strand breaks induced by fast and thermal neutrons from YAYOI research reactor in the presence and absence of boric acid, Radiat. Res. 191 (2019) 483–489.

[33] M.M. Vilenchik, A.G. Knudson, Endogenous DNA double-strand breaks: production, fidelity of repair, and induction of cancer, Proc. Natl. Acad. Sci. U.S.A. 100 (2003) 12871–12876.
[34] Y. Kinashi, S. Takahashi, G. Kashino, DNA double-strand break induction in Ku80-deficient CHO cells following boron neutron capture reaction, Radiat. Oncol. 6 (2011) 106.
[35] C. Rodriguez, M. Carpano, P. Curotto, In vitro studies of DNA damage and repair mechanisms induced by BNCT in a poorly differentiated thyroid carcinoma cell line, Radiat. Environ. Biophys. 57 (2018) 143–152.
[36] N. Kondo, H. Michiue, Y. Sakurai, Detection of γH2AX foci in mouse normal brain and brain tumor after boron neutron capture therapy, Rep. Pract. Oncol. Radiother. 21 (2016) 108–112.
[37] M. Masutani, D. Baiseitov, T. Itoh, Histological and biochemical analysis of DNA damage after BNCT in rat model, Appl. Radiat. Isot. 88 (2014) 104–108.
[38] N. Kondo, Y. Sakurai, Y. Hirota, DNA damage induced by boron neutron capture therapy is partially repaired by DNA ligase IV, Radiat. Environ. Biophys. 55 (2016) 89–94.
[39] K.H. Chen, Z.Y. Lai, D.Y. Li, Analysis of DNA damage responses after boric acid-mediated boron neutron capture therapy in hepatocellular carcinoma, Anticancer Res. 39 (2019) 6661–6671.
[40] O. Surova, B. Zhivotovsky, Various modes of cell death induced by DNA damage, Oncogene 32 (2013) 3789–3797.
[41] A. Takahashi, K. Ohnishi, I. Ota, p53-dependent thermal enhancement of cellular sensitivity in human squamous cell carcinomas in relation to LET, Int. J. Radiat. Biol. 77 (2001) 1043–1051.
[42] Y. Fujita, I. Kato, S. Iwai, Role of p53 mutation in the effect of boron neutron capture therapy on oral squamous cell carcinoma, Radiat. Oncol. 4 (2009) 63.
[43] K. Seki, Y. Kinashi, S. Takahashi, Influence of p53 status on the effects of boron neutron capture therapy in glioblastoma, Anticancer Res. 5 (2015) 169–174.
[44] N. Kondo, M. Hikida, M. Nakada, Glioma stem-like cells can be targeted in boron neutron capture therapy with boronophenylalanine, Cancers 12 (2020) 3040.
[45] T. Sun, Z. Zhang, B. Li, Boron neutron capture therapy induces cell cycle arrest and cell apoptosis of glioma stem/progenitor cells in vitro, Radiat. Oncol. 8 (2013) 195.
[46] H. Vakifahmetoglu, M. Olsson, B. Zhivotovsky, Death through a tragedy: mitotic catastrophe, Cell Death Differ. 15 (2008) 1153–1162.
[47] N.G. Oliveira, M. Castro, A.S. Rodrigues, Evaluation of the genotoxic effects of the boron neutron capture reaction in human melanoma cells using the cytokinesis block micronucleus assay, Mutagenesis 16 (2001) 369–375.
[48] S. Masunaga, K. Ono, Y. Sakurai, Evaluation of apoptosis and micronucleation induced by reactor neutron beams with two different cadmium ratios in total and quiescent cell populations within solid tumors, Int. J. Radiat. Oncol. Biol. Phys. 51 (2001) 828–839.
[49] T.E. Schmid, L. Canella, P. Kudejova, The effectiveness of the high-LET radiations from the boron neutron capture [^{10}B(n,a) ^{7} Li] reaction determined for induction of chromosome aberrations and apoptosis in lymphocytes of human blood samples, Environ. Biophys. 54 (2015) 91–102.
[50] Y. Fujita, N. Yamamoto, I. Kato, Induction of multinucleation in oral squamous cell carcinoma tissue with mutated p53 surviving boron neutron capture therapy, Int. J. Radiat. Biol. 87 (2011) 293–301.
[51] D. Ditsworth, W.X. Zong, C.B. Thompson, Activation of poly(ADP)-ribose polymerase (PARP-1) induces release of the pro-inflammatory mediator HMGB1 from the nucleus, J. Biol. Chem. 282 (2007) 17845–17854. Erratum in: J Biol Chem. 284 (2009):22500.

[52] S. Imamichi, L. Chen, T. Ito, Extracellular release of HMGB1 as an early potential biomarker for the therapeutic response in a xenograft model of boron neutron capture therapy, Biology (Basel) 11 (2022) 420.
[53] T.P. Lippert, R.A. Greenberg, The abscopal effect: a sense of DNA damage is in the air, J. Clin. Invest. 131 (2021) e148274.
[54] V.A. Trivillin, Y.V. Langle, M.A. Palmieri, Evaluation of local, regional and abscopal effects of Boron Neutron Capture Therapy (BNCT) combined with immunotherapy in an ectopic colon cancer model, Br. J. Radiol. 94 (2021) 20210593.
[55] V.A. Trivillin, E.C.C. Pozzi, L.L. Colombo, Abscopal effect of boron neutron capture therapy (BNCT): proof of principle in an experimental model of colon cancer, Radiat. Environ. Biophys. 56 (2017) 365–375.

CHAPTER SIX

Core level ionization or excitation and Auger relaxation induce clustered DNA damage

Akinari Yokoya[a,b,*] and Yui Obata[a,b]
[a]Institute for Quantum Life Science, National Institutes for Quantum Science and Technology, Chiba-shi, Japan
[b]Graduate School of Science and Engineering, Ibaraki University, Mito, Japan
*Corresponding author: e-mail address: yokoya.akinari@qst.go.jp

Contents

Abstract

Ionizing radiation causes various types of DNA damage, such as single- (SSBs) and double-strand breaks (DSBs), nucleobase lesions, abasic sites (AP sites), and cross-linking between complementary strands of DNA or DNA and proteins. DSBs are among the most harmful type of DNA damage, inducing serious genetic effects such as cell lethality and mutation. Nucleobase lesions and AP sites, on the other hand, may be less deleterious and are promptly repaired by base excision repair (BER) pathways. Recently, biochemical approaches to quantify nucleobase lesions and AP sites have revealed certain types of non-strand break lesions as harmful DNA damage, called clustered DNA

The Enzymes, Volume 51
ISSN 1874-6047
https://doi.org/10.1016/bs.enz.2022.08.006

damage. Such clusters can retard nucleobase excision repair enzymes, and can sometimes be converted to DSBs by BER catalysis. This unique character of clustered DNA damage strongly depends on the spatial density of ionization or excitation events occurring at the track end of initial radiation or low energy secondary electrons. In particular, the photoelectric effect of elements comprising biological molecules, followed by emission of Auger electrons, are key factors in determining the future fate of each clustered damage site. This chapter describes biological studies of clustered nucleobase lesions with SSBs or AP sites, and mechanistical studies on core level excitation and Auger relaxation giving rise to clustered DNA damage.

1. Introduction: Radiation damage to DNA

Ionizing radiation damages DNA in several ways, ultimately causing serious biological effects including cell lethality, mutation, senescence such as proliferation arrest, and inflammation or carcinogenesis (Fig. 1). Proximal strand breaks occur in both DNA complementary strands and are believed to be a major type of DNA damage (DSB), and many studies have attempted to quantify DSBs in cells exposed to various types of ionizing radiation [1]. These studies have shown that higher linear energy transfer (LET) radiation induces more DSBs than lower LET irradiation, indicating that severe biological effects are due to the biophysical nature of the dense ionization events caused by high-LET irradiation. Thus, the spatial structure of radiation is important in inducing DSBs.

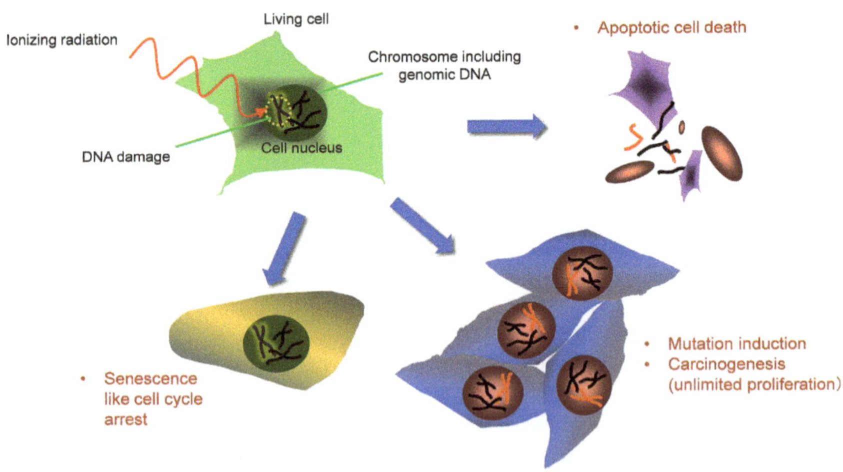

Fig. 1 Schematic illustration of the effects of ionizing radiation on living cells.

Living cells can normally repair most DSBs to avoid fatal damage of the genome through several pathways. In mammalian cells, non-homologous end joining or homologous recombination repair radiation-induced DSBs. Most DSBs (80–90%) initially produced in a cell are rejoined within several hours [2]. How cells choose the appropriate DSB repair pathway was reviewed [3,4]. The evidence strongly suggests that DSBs are mostly repaired, and radiobiological effects are induced by not only unrepaired DSBs but also the other types of genomic DNA damage. The detection of these latter types of damage is more challenging than DSBs, which result in changes in the molecular weight of DNA. Thus, early studies conducted computer simulations to predict the nature of "hidden" DNA damage not visualized or detected by conventional techniques.

2. Clustered DNA damage

Nikjoo et al. performed pioneer studies on the spatial distribution of isolated single-strand breaks (SSBs) or nucleobase lesions in DNA strands [5,6] using Monte Carlo track simulation methods. They revealed that the radiation tracks produce a cluster consisting of several independent proximal lesions. Ward and Goodhead named lesion clustering "multiple DNA damage sites" or "clustered DNA damage" [7,8]: two or more lesions (nucleobase lesion, SSB, abasic sites) formed within about 10 nucleobase pairs separated by a single radiation track. (In chapter "Carbon ion radiation and clustered DNA double-strand breaks": Fig. 1, Shibata describes the damage complexity in the proximity of DSBs.) Both damage clustering and DSBs are mainly caused by initial or secondary emitted low energy electrons and are thus categorized as "direct action" of radiation [9].

Most previous mechanistic studies focused on the indirect effects of radiation, by which DNA damage is induced by diffusible water radicals. These studies used dilute aqueous solutions containing DNA and showed that the hydroxyl radicals ($OH^{\bullet}$) is a major contributor of strand breaks [10], whereas both hydrated electrons and $OH^{\bullet}$ induce DNA nucleobase lesions [11–13]. Experimental [2] and theoretical [14] studies indicate that in living cells or under highly scavenging conditions similar to those involved in cellular $OH^{\bullet}$ scavenging, ~40% of the lesions induced in DNA by low-LET radiation could be ascribed to direct effects, increasing to ~70% for high-LET α-particles [15]. However, the mechanisms underlying the direct effects of radiation, particularly the induction of nucleobase lesions and AP sites in DNA by direct effects, are less well understood.

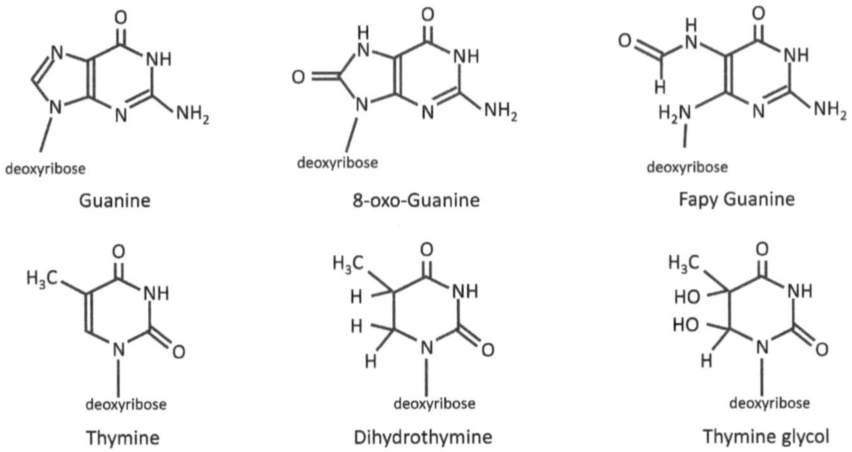

Fig. 2 Chemical structures of typical modified guanine and thymine nucleobases.

Nucleobase lesion induction has been studied by chromatographic analysis methods [16–18] and has revealed various oxidative or reductive modifications in purine or pyrimidine nucleobases. The major products resulting from electron loss processes are oxidized guanines, 2,6-diamino-4-hydroxy-5-*N*-methylformamidopyrimidine (Fapy), and 7,8-dihydro-8-oxo-2′-deoxyguanine (8-oxoG). The products arising from electron gain processes are ring-saturated pyrimidines (e.g., 5,6-dihydrothymine (DHT), thymine glycol, and abasic sites (AP sites)). The chemical structures of typical modified nucleobases are shown in Fig. 2. Chemical methodologies isolate nucleobase chemical species, and thus the positional relationship between each lesion in a DNA molecule was rarely identified. Alternative methods were required to validate the spatial distribution of lesions predicted by Monte Carlo track simulation, such as biochemical approaches using enzymatic probes.

3. Nucleobase excision repair enzymes as probes to detect nucleobase lesions and AP sites

Nucleobase excision repair pathways play an essential role to maintain fidelity in DNA replication against external mutagenic stresses. The main players of these pathways are several base excision repair (BER) enzymes, which are very conservatively encoded in the genomes of prokaryotes and eukaryotes [19]. First, the BER enzymes recognize nucleobase lesions and excise the damaged nucleobases by the nucleobase excision activity of glycosylases. Next, most BER enzymes snip at the AP site in the DNA backbone resulting from nucleobase excision, causing nicks (SSBs). Endonuclease III (EndoIII, or NTH) in *Escherichia coli* is a well-characterized nucleobase excision enzyme that mainly excises pyrimidine damage such as DHT, Tg, and AP sites [20–23]. Human cells possess an analogue of this protein (e.g., hNTH). Formamidopyrimidine-DNA glycosylase (FPG) from *E. coli* excises purine lesions (mainly 8-oxoG) and AP sites [24–26]. In human cells, hOGG1 (human oxoguanine glycosylase 1) is a functional analogue of FPG. These BER proteins are thought to be enzymatic probes which convert nucleobase lesions or AP sites to readily detectable DNA strand breaks, indicating that enzymatic treatment could be an alternative method of detecting nucleobase lesions in DNA in vitro or in a cell nucleus.

Dizdaroglu et al. [27] were the first to utilize the enzymatic function of NTH, and revealed the enzymatic substrates produced by ionizing irradiation and investigated the degree nucleobase damage using NTH. They used plasmid DNA to detect nucleobase lesions induced by exposure to α-particles or γ-rays. The nucleobase lesions were converted to SSBs by the glycosylase and AP lyase activities of the BER enzymes, and visualized as open circular or linear forms of the plasmid using agarose gel electrophoresis (Fig. 3). Thereafter, many studies explored nucleobase lesions caused by various types of radiation using NTH, FPG or endonuclease IV (NFO), another BER enzyme that recognizes AP sites.

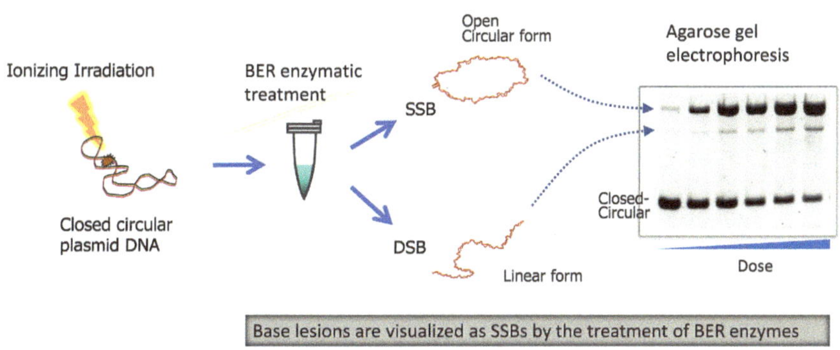

Fig. 3 Basic experimental procedure for the visualization of nucleobase lesions produced in plasmid DNA induced by irradiation. Conformational changes of the plasmid DNA induced by treatment with base excision repair enzyme are analyzed using agarose gel electrophoresis.

4. DSBs arising from treatment with base excision enzymes

Many studies using BER enzymes acting on nucleobase damage have greatly increased our understanding of non-DSB-type clustered DNA damage caused by exposure to ionizing radiation. Conversion of a nucleobase lesion(s) into a SSB by BER protein probes results in a DSB if the two lesions are proximately located within 10 nucleobase pairs of each other, one on each strand [28]. The use of a simple model DNA, such as plasmid DNA, produces enzymatically induced DSBs which indicate that the nucleobase lesion(s) is produced together with another nucleobase lesion(s), AP site(s) or SSB(s). As shown in Fig. 3, in addition to DSBs formed immediately by exposure to irradiation, non-DSB-type clustered DNA damage is observed as additional DSBs in linear form DNA by agarose gel electrophoresis. Sutherland and coauthors used the enzymatic probes FPG and NFO to further complex genomic DNA samples from mammalian cells. After enzymatic treatment of large genomic DNA extracted from cultured cells exposed to X-rays or γ-rays, they analyzed the sample by pulse-field gel electrophoresis and found that the enzymatic probes induced a significant level of additional DSBs (Fig. 4, [29,30]). This suggests that DSBs visualized in the irradiated cells might include DSBs additionally produced by BER proteins

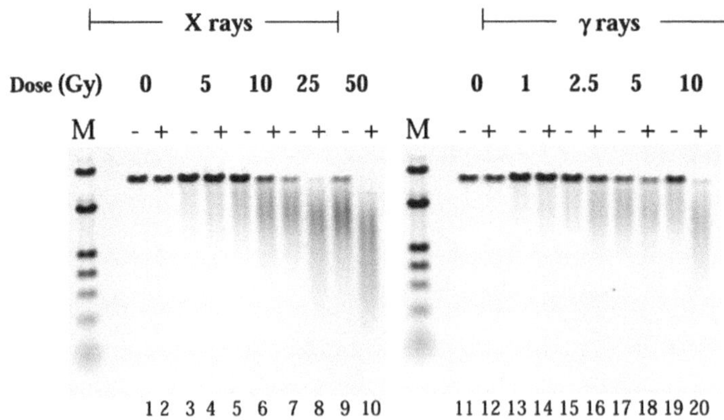

Fig. 4 Image of a pulsed-field agarose gel for analyzing damage cluster produced in T7 phage DNA exposed to 50-kVp X-rays (left panel) and T7 phage DNA exposed to ^{137}Cs γ-rays. *The image is reproduced from B.M. Sutherland, P.V. Bennett, O. Sidorkina, J. Laval, Clustered DNA damages induced in isolated DNA and in human cells by low doses of ionizing radiation, Proc. Natl. Acad. Sci. U.S.A. 97 (2000) 103–108, with permission.*

because catalytic reactions of these enzymes are mostly complete within a few hours before microscopic observation of the cells or biochemical analysis. In other words, the number of DSBs reported in previous studies on irradiated cells might be overestimations.

To date, most studies have revealed that ionizing radiation causes not only additional SSBs but also additional DSBs, as determined by BER enzymatic treatment. The evidence demonstrates that significant amounts of nucleobase lesions or AP sites are produced by ionizing irradiation. Both γ-rays [31] and X-rays [32], as well as high LET ion beams [33], create approximately three times the number of nucleobase lesions (or AP sites), and their clusters. Previously, these low-LET radiations were believed to cause less deleterious DNA damage (such as DSBs) as compared with high-LET radiation such as α-particles or ion beams from an accelerator. The mechanism of damage clustering by low-LET irradiation has not been fully elucidated. The effects of dense ionization or excitation events along the low energy secondary electron tracks, including Auger electrons, may result in proximal lesions within a region of DNA several nanometers long [34]. These low energy electrons originate from photoelectric effects of the DNA constituent atoms: hydrogen (H), carbon (C), nitrogen (N), oxygen (O) and phosphorus (P). It should be emphasized that these effects are caused not only by low-LET irradiation, but also by collisions with various charged particles through electrostatic interaction, as described in the following sections.

5. Core level ionization or excitation by collision with swiftly moving high-LET charged particles or irradiation with low-LET radiations

In a landmark paper, Platzman [35] suggested that the ejection of inner-shell electrons, especially K-shell electrons in carbon, nitrogen, or oxygen atoms, by collision with swiftly moving charged particles followed by Auger transition may result in massive energy transfer to a biological molecule. His suggestion at a symposium on radiobiology in 1950 preceded the publication of the double helix model of DNA by Watson and Crick [36] by 3 years. Although the fraction of energy thus transferred in collisions to inner-shell electrons is a few percent or less compared with valence electron ionization, Platzman concluded that this transfer is not negligible and might, in some instances, play a significant role in radiobiological effects.

Chetioui and colleagues [37] performed pioneering theoretical work on ion-particle irradiation, such as helium or ion atoms with various LET values, of biological materials. They calculated the cross-sections of K-shell ionizations of carbon, oxygen, and phosphorus atoms based on the first Born approximation. They found that the LET dependence of cross sections of K-shell electron ionization by collision with projectile ions resulting a K-shell vacancy was consistent with the LET-dependence of the experimentally reported cell-lethal cross-sections. Interestingly, the cross-sections of carbon and oxygen atoms, rather than that of phosphorus atoms, contributed significantly to this dependency.

Core ionization and subsequent Auger relaxation cause emission of a substantial number of Auger electrons as well as a photoelectron from these atoms. An inner or outer shell electron absorbing a photon is ejected from the atom as a photoelectron, whose energy is determined by the incident X-ray photon energy and electron binding energy in the atom. When an inner shell electron is ejected, Auger relaxation may be induced to compensate the higher excitation state of the atom by ejecting other electrons in the atom as Auger electrons. The energy of the Auger electrons is coincident with the energy difference between two particular electronic states.

The accompanying impacts of Auger electrons may eventually cause various types of damage in genomic DNA. These Auger effects can also be induced in biological material by irradiation with conventional ionizing radiation through the impact of energetic (secondary) electrons. Bernhardt et al. [38,39] simulated the contribution of low energy Auger electrons in biological effects, such as cell lethality, using the Monte Carlo track structure simulating program PARTRAC. They suggested that, in the 0.3–0.5 keV photon (X-ray) energy range, highly efficient biological effects could be due to an increased local dose deposited by low energy Auger electrons to DNA. Boissière et al. [40] calculated the number of K-ionization events in a V79 Chinese hamster cell nucleus exposed to low LET γ-rays or to 100 keV electrons, and suggested that these Auger electrons would contribute about 75% of V79 cell lethality.

Mechanistic studies have also been performed using in vitro model DNA molecules. X-rays with energies less than several keV are called soft X-rays and are suitable probes for generating substantial numbers of Auger electrons through a photoelectric effect. An inner or outer shell electron absorbing a photon is ejected from the atom as a photoelectron, whose energy is determined by the incident X-ray photon energy and the electron binding energy in the atom (Fig. 5). When a K-shell electron in a DNA constituent atom is

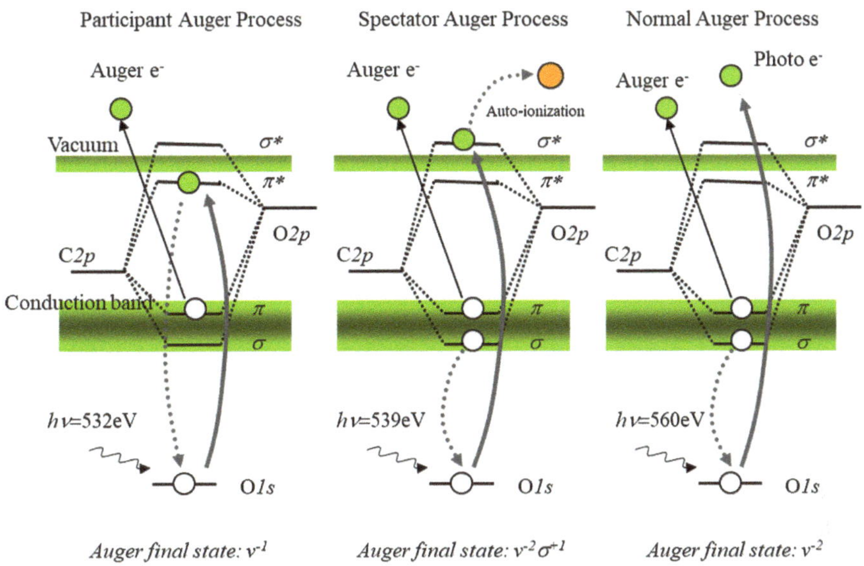

Fig. 5 Schematic illustration of three types of Auger relaxation, and emission of Auger electrons.

ionized or excited into an antibonding orbital just above the ionization potential, the kinetic energy of the emitted electrons as a consequence of Auger relaxation is typically several hundred eV or less. These electrons strongly interact with the other atoms through inelastic scattering and lose energy, and finally form hydrated electrons in aqueous solution. The mean free path of these electrons is less than a few tens of nanometers. The electrons deposit their energy within macromolecules such as DNA and produce very localized secondary ionizations or electron excitations. Thus, Auger electrons are likely to cause clustered DNA damage sites in living cells.

Watanabe et al. [41] performed Monte Carlo track simulations of Auger and photo-electrons from C, N, O and P atoms in DNA and estimated the yields of SSBs and DSBs. They assumed that the strand breaks were induced by energy deposition into DNA simply by K-shell ionization and emitted photo- or Auger electrons in DNA (model 1), or that energy deposition into the hydrated layer also contributed to strand breaks (model 2). As shown Fig. 6, the yields of strand breaks were enhanced by nitrogen (400 eV), oxygen (530 eV) and phosphorus (2100 eV) K-shell ionization. In the figure, the ratio of SSBs to DSBs (SSB/DSB), or the ratio of complex DSBs with additional SSBs around the termini to simple DSBs, also showed that the Auger relaxations of these elements enhance the complexity of DNA

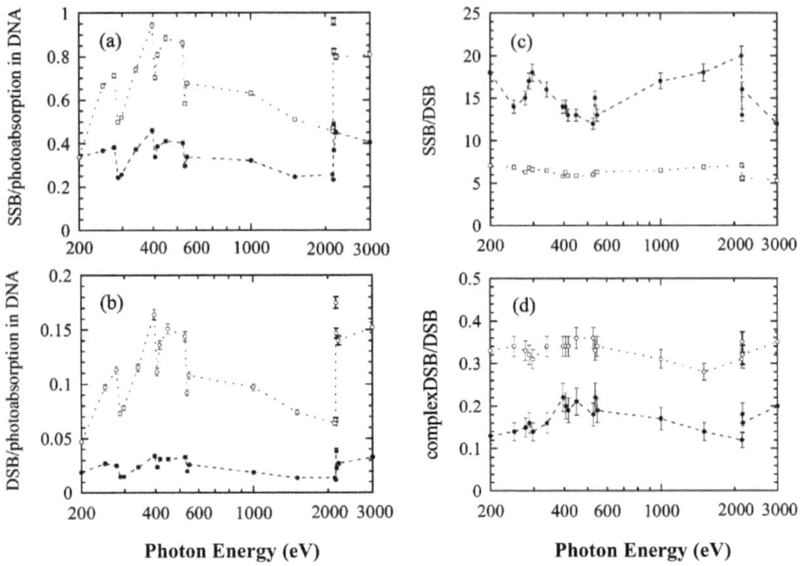

Fig. 6 Yields of SSBs (panel A), DSBs (panel b), the ratios of SSBs to DSBs (panel C) and ratios of complex DSBs to total DSBs (d) as a function of soft X-ray energy. *The graphs are modified from R. Watanabe, A. Yokoya, K. Fujii, K. Saito, DNA strand breaks by direct energy deposition by Auger and photo-electrons ejected from DNA constituent atoms following K-shell photoabsorption, Int. J. Radiat. Biol. 80 (2004) 823–832, with permission.*

damage. Although many studies report strand break induction in this soft X-ray energy region [42–44], less was known until recently about the induction of nucleobase lesions and clustered damage in DNA caused by Auger relaxation.

6. Synchrotron radiation providing monochromatic soft X-rays induce Auger relaxation of particular elements in DNA

Synchrotron radiation facilities have been used as intense sources of hard X-rays (~100 keV) through to infrared (IR, below 1 eV). A pioneer study of the biological effects of synchrotron radiation in 1976 was conducted at the Synchrotron Orbital Radiation Facility (SOR) at the University of Tokyo (Tanashi, Tokyo, Japan) [45]. Since then, many studies have focused on Auger enhancement of various biological effects (e.g., DNA strand breaks, cell lethality, mutation induction, chromosomal aberration) [46]. Typical DNA damage, such as strand breaks, was determined using a model DNA molecule such as plasmid exposed to synchrotron X-rays highly monochromatized by a grating or double-crystal monochromator, and the results were analyzed by gel electrophoresis.

In a pioneer study of DNA damage caused by core-ionization, Fayard and colleagues investigated whether DNA strand breaks were enhanced by carbon K-shell ionization [47]. They examined the strand break yields produced in dry pBS plasmid DNA exposed to monochromatic soft X-rays using the Laboratoire pour l'Utilisation du Rayonnement Electromagnétique (LURE) synchrotron facility in Orsay, France. The energies used (250 or 380 eV) were below and above the carbon K-ionization threshold (278 eV), respectively. They found that the yield of SSBs was slightly enhanced, and that of DSB significantly enhanced, by carbon K-ionization. To access a higher energy region encompassing the nitrogen and oxygen K-shell ionization thresholds, Yokoya et al. [43] used the Photon Factory (PF) of the High Energy Accelerator Research Organization (KEK), and found that the yields of SSBs and DSBs in a dried plasmid DNA film in a vacuum chamber exposed to radiation were constant regardless of the photon energy irradiated. They suggested that K-ionization followed by Auger relaxation could induce DNA strand breaks with constant efficiency in terms of the absorbed energy. Similar constant strand break yields of pBR322 plasmid DNA were reported using phosphorus K-ionization soft X-rays provided by the PF [48]. Using LURE soft X-rays, Eschenbrenner et al. investigated the strand break yields of hydrated pBR or pSP plasmid films without any buffer solutes [49]. They found that the efficiencies of DSBs induced by oxygen K-ionization at 760 eV was greater than those by carbon K-ionization at 380 eV, suggesting that the hydration level dependence of strand break yields may be associated with charge transfer from the hydration shell to the DNA. The photons above the oxygen K-threshold at 760 eV are more often absorbed in the hydration shell, and thus induce more DSBs.

Agrawala et al. [50] reported similar results using hydrated pBR or pSP plasmid films with buffer solutes. These early studies indicated that the yields of strand breaks depend not only on the K-shell ionized elements which give various ranges of secondary electrons, but also on the hydration level of the sample. Strand breaks in DNA are caused by the scission of chemical bond(s) between atoms, of which one is core-ionized. Ejected secondary electrons from K-ionized atoms (photo-electrons or Auger electrons) resulting from the photoelectric effect can also cause strand breaks with roughly similar efficiency because the energy spectra of these electrons are similar [41], ranging from tens to hundreds of eV.

Using FPG protein as a probe to excise a purine base lesion such as 8-oxoG, Agrawala et al. [50] quantified purine lesions for the first time, as well as strand breaks produced in hydrated pBR or pSP plasmid films exposed to 250, 380 and 760 eV soft X-rays. Contrary to expectation, no nucleobase lesions sensitive to FPG were enhanced high above the oxygen K-threshold. Fujii et al. [51] similarly reported that yields of FPG sensitive sites produced in a dry pUC18 plasmid DNA film exposed to 760 eV soft X-rays at SPring-8 were similar to those generated by exposure to 270, 380 and 435 eV photons (Fig. 7) [47]. Interestingly, both FPG and NTH sensitive sites were significantly enhanced by irradiation by 560 eV soft X-ray. Although the exposures to 760 eV photons tested by the two groups are considerably higher than the oxygen K-threshold (530 eV), 560 eV photons (just above the oxygen K-ionization threshold) are likely efficiently absorbed by oxygen atoms in DNA and by the hydrating 2.5 water molecules tightly bound to a nucleotide under high vacuum [52], resulting in nucleobase lesions. Additional DSBs generated by enzymatic treatment also showed slightly higher yields when compared with yields following exposure to other soft X-ray energies [50]. These studies showed that oxygen and some carbon K-ionizations, followed by Auger relaxation, but not nitrogen K-ionization in the nucleobases, likely contribute to the induction of clustered DNA damage sites, including nucleobase lesions.

7. Model of clustered damage induced after Auger relaxation of DNA constituent atoms

The yields of NTH sensitive sites are significantly larger than those of FPG sensitive sites at 760 eV. Many studies report larger yields for dry or hydrated DNA film samples treated with NTH after exposed to γ-irradiation [31], 150 kVp X-rays [32], He^{2+} ions with various LET values [53], and other ion beams (He^{2+}, $C^{5+,6+}$ and $Ne^{8+,10+}$) [33]. Interestingly, the greater sensitivity of NTH sites compared to FPG sensitive sites decreases in solution in association with radical scavenging capacity: in a diluted solution, the nucleobase damage yields for both enzymatic treatments are similar, whereas the fraction of FPG sensitive sites largely decreases with increasing radical scavenging capacity [54]. Similarly, the relative fraction of AP sites revealed by NFO proteins as compared with NTH sensitive sites decreased with increasing scavenging capacity [54]. Hence, purine nucleobase damage would be mainly caused by water radiolysis radicals and purine damage by direct energy deposition of photo- or Auger electrons produced at other

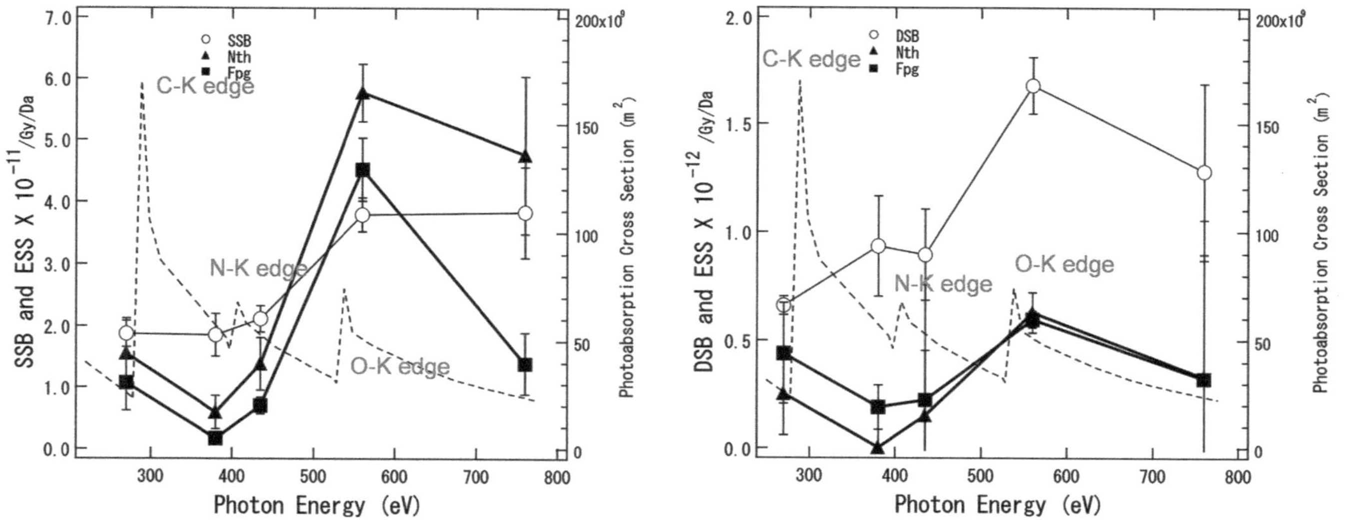

Fig. 7 Soft X-ray energy dependence of the yield of SSBs (left panel) and DSBs (right panel) produced in dry plasmid (pUC18) DNA thin films irradiated with 270, 380, 435, 560, and 760 eV at room temperature (○) or postirradiation treatment with NTH (▲) or FPG (■). Solid lines are drawn by connecting the data points. The dashed lines show the photoabsorption spectra of DNA for comparison. *The figures are modified from the original figure in K. Fujii, N. Shikazono, A. Yokoya, Nucleobase lesions and strand breaks in dry DNA thin film selectively induced by monochromatic soft X-rays, J. Phys. Chem. B. 113 (2009) 16007–16015, with permission.*

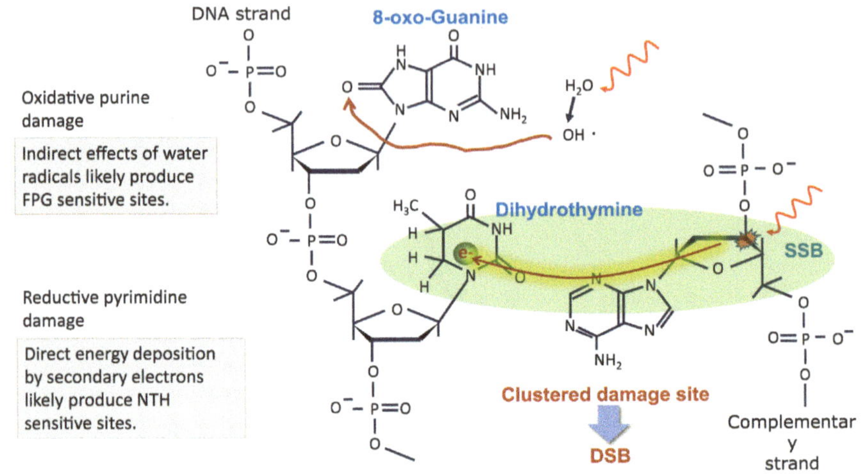

Fig. 8 Model of the induction of nucleobase lesions, SSBs, and their clustering.

sites of the DNA molecule. When a photoabsorption site (which is presumably converted to a SSB site) locates proximal to a pyrimidine lesion, it may be converted to a DSB after NTH treatment (Fig. 8).

8. Pre-thermalized Auger electrons are highly localized by the Coulomb potential of the cation ion of the parent atom

When Auger or photoelectrons are ejected from their parent atom, they move in the Coulomb potential of the cation, suggesting that their range is strongly restricted. Until recently, however, Monte Carlo track simulation studies have not taken into account the electrostatic effect of cations. Kai et al. conducted dynamic Monte Carlo simulations of electron trajectories in the Coulomb potential formed in liquid water to solve Newton's equation [55]. To better estimate the electron range, they also calculated cross sections, stopping powers, and energy loss rates for rotational and phonon excitation processes in liquid water by electron impact [56]. Using these physical parameters, they revealed the electron dynamics of pre-thermalized electrons before forming completely hydrated electrons, and compared the loci of ionization events, or electric or phonon excitation events, with the scale of a DNA molecule to predict the DNA damage localization mechanism [57]. They focused on the spatial and temporal distributions of Auger electrons or photoelectrons, and the collision events (e.g., ionization, electronic excitation, and dissociative electron attachment) involved in DNA damage localization. These low energy secondary electrons (below several hundred eV) are decelerated in water by the Coulombic force and to some extent recombine to ionized atoms (cations). Furthermore, some considerably decelerated electrons might be pre-hydrated in the water layer near the parent cation in DNA if the electrons are emitted from the DNA. Pre-hydrated electrons generated by return of an electron might play a significant role in forming a damage cluster site [58,59] (Fig. 9).

9. Novel method to examine the repairability of various types of DNA damage, and future prospects of studies on Auger relaxation and clustered DNA damage

Several in vitro studies have shown the number of AP sites in an AP cluster using FRET detection [60], or by visualizing nucleobase lesions and complex damage using atomic force microscopy after treating the damaged sites by aldehyde reactive probe (ARP) and ARP-binding protein [61]. However, less is known about the repairability of DNA damage, including clustered damage, in living cells, especially in terms of repair dynamics. Previous studies using live cells showed that whole-cell irradiation causes damage not only to genomic DNA but also to cytoplasmic components. Many studies using microbeams revealed that the latter process also significantly affects cellular viability [62,63]. However, it is difficult to completely isolate the DNA repair process from cytoplasmic effects ([63] and references therein] because cytoplasmic effects are induced by narrow radiation beams which penetrate the cytoplasm.

Transfection of irradiated DNA into nonirradiated cells was suggested as a method to realize the complete set of cytoplasmic functions. In early studies focusing solely on the cellular response to DNA damage, Thacker et al. [64] and Rainbow et al. [65] infected nonirradiated human cells with γ-ray irradiated human adenovirus. They studied SSB repair efficiency in the adenovirus genome by analyzing the molecular weight of the recovered DNA following each treatment. Nakaue et al. [66] recently established a novel experimental system consisting of an X-ray-irradiated plasmid expressing enhanced green fluorescent protein (EGFP) transfected into nonirradiated human cells (Fig. 10) to investigate the efficiency of repair of radiation-induced non-DSB DNA damage. Using this system, Obata et al. [67] revealed that the SSBs produced in plasmids exposed to conventional X-rays were less repairable as compared to strand break sites produced by nicking enzymes (Fig. 11). Rejoining of the radiation-induced strand break termini can be slower than by enzymatic treatment, suggesting that the assay can highly sensitively detect the repairability of DNA damage. Complex

(a)

(b)

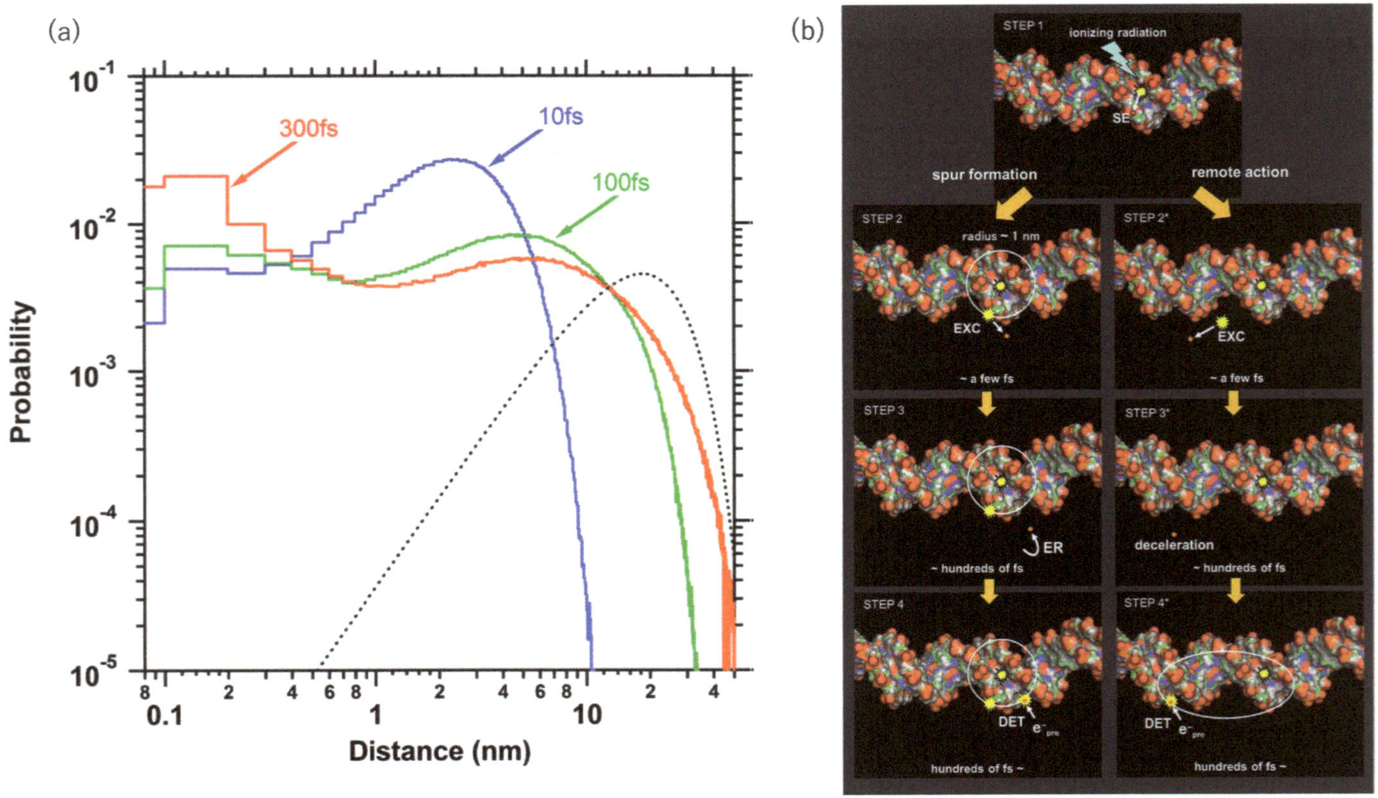

Fig. 9 Temporal evolution of spatial probability distributions of secondary electrons as a function of distance from the ionization site. The dotted line shows the results obtained using a previous method (R. H. Ritchie et al. in Computational Approaches in Molecular Radiation Biology (Basic Life Sciences 63); Varma, M. N., Chatterjee, A, Eds., Plenum Press, New York, 1994; pp. 33–44).

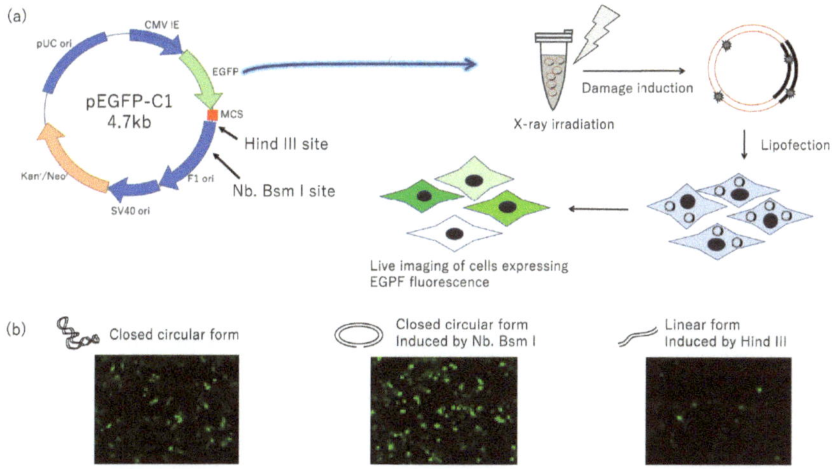

Fig. 10 Schematic of the experiment involving X-ray irradiation of the EGFP expressing plasmid (pEGFP-C1), lipofection into human cells (MCF7), and live-cell observation of green fluorescence. *The figure is modified from the original figure in Y. Obata, H. Nakaue, K. Hirasaki, N. Akimitsu, A. Yokoya, Expression of an X-irradiated EGFP-expressing plasmid transfected into non-irradiated human cells, Radiat. Res. 196 (2021) 261–271, with permission.*

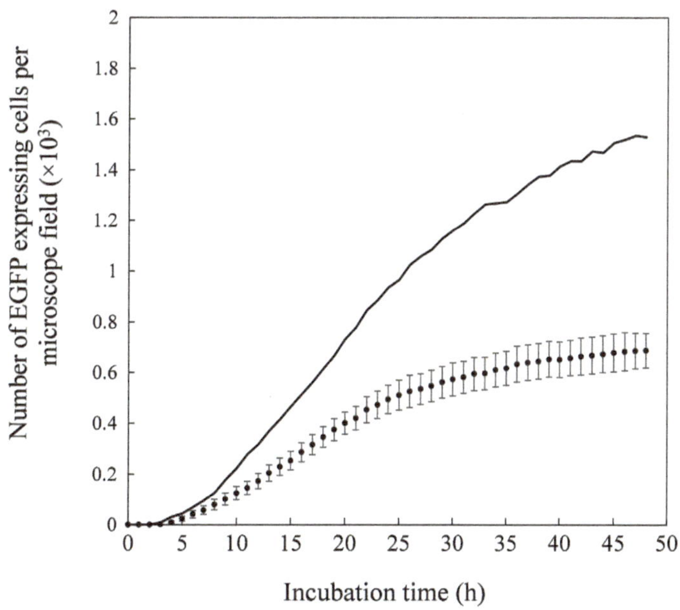

Fig. 11 Comparison of the calculated number of green fluorescent (EGFP-expressing) MCF7 cells to the corresponding experimental data. The calculated values (solid line) were obtained based on Poisson statistics assuming that radiation strikes the plasmid randomly. The dotted line indicates the experimental values. *This figure is modified from the original figure in Y. Obata, H. Nakaue, K. Hirasaki, N. Akimitsu, A. Yokoya, Expression of an X-irradiated EGFP-expressing plasmid transfected into non-irradiated human cells, Radiat. Res. 196 (2021) 261–271, with permission.*

strand break termini with nucleobase lesions are detected as decreased expression of the fluorescent protein. In future studies, such approaches could contribute to elucidating the repairability of clustered DNA damage caused by various core level ionizations or excitations, followed by Auger relaxation emitting low energy secondary electrons.

10. Conclusion

Core level ionization or the excitation of DNA constituent element atoms likely causes clustered DNA damage, along or at the end of various radiation tracks. These clustered damage sites are associated with oxidative purine lesions mainly induced by water radicals (indirect effect) or reductive pyrimidine lesions as a result of reactions with low energy Auger or photoelectrons. The secondary electrons experience an attractive force from their parent cation and spatially and temporally localize around their origin, resulting in a cluster of lesions. The repairability of these clustered damage sites is currently being elucidated. The transfection of various functional plasmids as models of damaged DNA into non-irradiated cells is a promising technique for future studies.

References

[1] K.M. Prise, G. Ahnström, M. Belli, J. Carlsson, D. Frankenberg, J. Kiefer, M. Löbrich, B.D. Michael, J. Nygren, G. Simone, B. Stenerlöw, A review of dsb induction data for varying quality radiations, Int. J. Radiat. Biol. 74 (1998) 173–184.
[2] C.M. deLara, T.J. Jenner, K.M. Townsend, S.J. Marsden, P. O'Neill, The effect of dimethyl sulfoxide on the induction of DNA double-strand breaks in V79-4 mammalian cells by alpha particles, Radiat. Res. 144 (1995) 43–49.
[3] R. Scully, A. Panday, R. Elango, N.A. Willis, DNA double-strand break repair-pathway choice in somatic mammalian cells, Nat. Rev. Mol. Cell Biol. 20 (2019) 698–714.
[4] A. Shibata, Regulation of repair pathway choice at two-ended DNA double-strand breaks, Mutat. Res. 803-805 (2017) 51–55.
[5] H. Nikjoo, D.T. Goodhead, D.E. Charlton, H.G. Paretzke, Energy deposition in small cylindrical targets by ultrasoft x-rays, Phys. Med. Biol. 34 (1989) 691–705.
[6] D.T. Goodhead, H. Nikjoo, Track structure analysis of ultrasoft X-rays compared to high- and low-LET radiations, Int. J. Radiat. Biol. 55 (1989) 513–529.
[7] J.F. Ward, DNA damage produced by ionizing-radiation in mammalian-cells—identities, mechanisms of formation, and reparability, Prog. Nucleic Acid Res. Mol. Biol. 35 (1988) 95–125.
[8] D.T. Goodhead, Initial events in the cellular effects of ionizing radiations: clustered damage in DNA, Int. J. Radiat. Biol. 65 (1994) 7–17.
[9] W.A. Bernhard, D.M. Close, DNA damage dictates the biological consequences of ionizing irradiation: the chemical pathways, in: A. Mozumder, Y. Hatano (Eds.), Charged Particle and Photon Interactions with Matter, Marcel Dekker, New York, 2003, pp. 471–489.

[10] P. O'neill, E.M. Fielden, Primary free radical processes in DNA, Adv. Radiat. Biol. 17 (1993) 53–120.
[11] C. von Sonntag, The Chemical Basis of Radiation Biology, Taylor & Francis, New York, 1987.
[12] A.F. Fuciarell, B.J. Wegher, E. Gajewski, M. Dizdaroglu, W.F. Blakel, Quantitative measurement of radiation-induced base products in DNA using gas chromatography-mass spectrometry, Radiat. Res. 119 (1989) 219–231.
[13] P. O'Neill, Radiation-induced damage in DNA, in: Radiation Chemistry, Elsevier, Science, Dordrecht, 2001, pp. 585–622.
[14] H. Nikjoo, C.E. Bolton, R. Watanabe, M. Terrissol, P. O'Neill, D.T. Goodhead, Modelling of DNA damage induced by energetic electrons (100 eV to 100 keV), Radiat. Prot. Dosimetry 99 (2002) 77–80.
[15] D. Becker, M.D. Sevilla, The chemical consequences of radiation damage to DNA, Adv. Radiat. Biol. 17 (1993) 121–180.
[16] M. Dizdaroglu, Application of capillary gas chromatography-mass spectrometry to chemical characterization of radiation-induced base damage of DNA: implications for assessing DNA repair processes, Anal. Biochem. 144 (1985) 593–603.
[17] T. Mori, Y. Hori, M. Dizdaroglu, DNA base damage generated in vivo in hepatic chromatin of mice upon whole body gamma-irradiation, Int. J. Radiat. Biol. 64 (1993) 645–650.
[18] J. Cadet, T. DoukiJ, L. Ravanat, Artifacts associated with the measurement of oxidized DNA bases, Environ. Health Perspect. 105 (1997) 1034–1039.
[19] H. Ide, DNA substrates containing defined oxidative base lesions and their application to study substrate specificities of base excision repair enzymes, Prog. Nucleic Acid Res. Mol. Biol. 68 (2001) 207–221.
[20] B. Demple, S. Linn, DNA N-glycosylases and UV repair, Nature 287 (1980) 203–208.
[21] L.H. Breimer, T. Lindahl, DNA glycosylase activities for thymine residues damaged by ring saturation, fragmentation, or ring contraction are functions of endonuclease III in Escherichia coli, J. Biol. Chem. 259 (1984) 5543–5548.
[22] K.M. Prise, C.H.L. Pullar, B.D. Michael, A study of endonuclease III-sensitive sites in irradiated DNA: detection of α-particle-induced oxidative damage, Carcinogenesis 20 (1999) 905–909.
[23] Z. Hatahet, Y.W. Know, A.A. Purmal, R.P. Cunningham, S.S. Wallace, New substrates for old enzymes. 5-Hydroxy-2′-deoxycytidine and 5-hydroxy-2′-deoxyuridine are substrates for Escherichia coli endonuclease III and formamidopyrimidine DNA N-glycosylase, while 5-hydroxy-2′-deoxyuridine is a substrate for uracil DNA N-glycosylase, J. Biol. Chem. 269 (1994) 18814–18820.
[24] C.J. Chetsanga, T. Lindahl, Release of 7-methylguanine residues whose imidazole rings have been opened from damaged DNA by a DNA glycosylase from *Escherichia coli*, Nucleic Acids Res. 6 (1979) 3673–3684.
[25] S. Boiteux, E. Gajewski, J. Laval, M. Dizdaroglu, Substrate specificity of the *Escherichia coli* Fpg protein (formamidopyrimidine-DNA glycosylase): excision of purine lesions in DNA produced by ionizing radiation or photosensitization, Biochemistry 31 (1992) 106–110.
[26] J. Tchou, H. Kasai, S. Shibutani, M.H. Chung, J. Laval, A.P. Grollman, S. Nishimura, 8-oxoguanine (8-hydroxyguanine) DNA glycosylase and its substrate specificity, Proc. Natl. Acad. Sci. U. S. A. 88 (1991) 4690–4694.
[27] M. Dizdaroglu, J. Laval, S. Boiteux, Substrate specificity of the *Escherichia coli* endonuclease III: excision of thymine- and cytosine-derived lesions in DNA produced by radiation-generated free radicals, Biochemistry 32 (1993) 12105–12111.
[28] R. Hanai, M. Yazu, K. Hieda, On the experimental distinction between ssbs and dsbs in circular DNA, Int. J. Radiat. Biol. 73 (1998) 475–479.

[29] B.M. Sutherland, P.V. Bennett, O. Sidorkina, J. Laval, Clustered DNA damages induced in isolated DNA and in human cells by low doses of ionizing radiation, Proc. Natl. Acad. Sci. U. S. A. 97 (2000) 103–108.
[30] B.M. Sutherland, P.V. Bennett, O. Sidorkina, J. Laval, Clustered damages and total lesions induced in DNA by ionizing radiation: oxidized bases and strand breaks, Biochemistry 39 (2000) 8026–8803.
[31] A. Yokoya, S.M.T. Cunniffe, P. O'Neill, Effect of hydration on the induction of strand breaks and base lesions in plasmid DNA films by gamma-radiation, J. Am. Chem. Soc. 124 (2002) 8859–8866.
[32] A. Yokoya, K. Fujii, T. Ushigome, N. Shikazono, A. Urushibara, R. Watanabe, Yields of strand breaks and base lesions induced by soft X-rays in plasmid DNA, Radiat. Prot. Dosimetry 122 (2006) 86–88.
[33] T. Ushigome, N. Shikazono, K. Fujii, R. Watanabe, M. Suzuki, C. Tsuruoka, H. Tauchi, A. Yokoya, Yield of single- and double-strand breaks and nucleobase lesions in fully hydrated plasmid DNA films irradiated with high-LET charged particles, Radiat. Res. 177 (2012) 614–627.
[34] H. Nikjoo, P. O'Neill, W.E. Wilson, D.T. Goodhead, Computational approach for determining the spectrum of DNA damage induced by ionizing radiation, Radiat. Res. 156 (2001) 577–583.
[35] R.L. Platzman, On the primary processes in radiation chemistry and biology, in: J.J. Nickson, P. Morrison, M. Burton, E.S. Guzman Barron, G. Failla, H.T. Patt (Eds.), Symposium on Radiobiology, the Basic Aspects of Radiation Effects on Living Systems, John Wiley and Sons Inc, New York, 1952, pp. 97–116.
[36] J.D. Watson, F.H. Crick, Molecular structure of nucleic acids; a structure for deoxyribose nucleic acid, Nature 171 (1953) 737–738.
[37] A. Chetioui, I. Despiney, L. Guiraud, L. Adoui, L. Sabatier, B. Dutrillaux, Possible role of inner-shell ionization phenomena in cell inactivation by heavy ions, Int. J. Radiat. Biol. 65 (1994) 511–522.
[38] P. Bernhardt, W. Friedland, P. Jacob, H.G. Paretzke, Modeling of ultrasoft X-ray induced DNA damage using structured higher order DNA targets, Int. J. Mass Spectrom 223 (224) (2003) 579–597.
[39] P. Bernhardt, W. Friedland, H.G. Paretzke, The role of atomic inner shell relaxations for photon-induced DNA damage, Radiat. Environ. Biophys. 43 (2004) 77–84.
[40] A. Boissière, C. Champion, A. Touati, M.A. Hervé du Penhoat, L. Sabatier, A. Chatterjee, A. Chetioui, DNA core ionization and cell inactivation, Radiat. Res. 167 (2007) 493–500.
[41] R. Watanabe, A. Yokoya, K. Fujii, K. Saito, DNA strand breaks by direct energy deposition by Auger and photo-electrons ejected from DNA constituent atoms following K-shell photoabsorption, Int. J. Radiat. Biol. 80 (2004) 823–832.
[42] K. Hieda, T. Hirono, A. Azami, M. Suzuki, Y. Furusawa, H. Maezawa, N. Usami, A. Yokoya, K. Kobayashi, Single- and double-strand breaks in pBR322plasmid DNA by monochromatic X-rays on and off the K-absorption peak of phosphorus, Int. J. Radiat. Biol. 70 (1996) 437–445.
[43] A. Yokoya, R. Watanabe, T. Hara, Single- and double-strand breaks in solid pBR322 DNA induced by ultrasoft X-rays at photon energies of 388, 435 and 573 eV, J. Radiat. Res. 40 (1999) 145–158.
[44] B. Fayard, A. Touati, F. Abel, M.A. Hervé du Penhoat, I. Despiney-Bailly, F. Gobert, M. Ricoul, A. L'Hoir, M.F. Politis, M.A. Hill, D.L. Stevens, L. Sabatier, E. Sage, D.T. Goodhead, A. Chetioui, Cell inactivation and double-strand breaks: the role of core ionizations, as probed by ultrasoft X rays, Radiat. Res. 157 (2002) 128–140.
[45] T. Ito, A. Ito, Predominance of membrane damage in yeast cells in suspension with monochromatic 163-nm vacuum ultraviolet light, Radiat. Res. 81 (1980) 416–426.

[46] A. Yokoya, T. Ito, Photon-induced Auger effect in biological systems: a review, Int. J. Radiat. Biol. 93 (2017) 743–756.
[47] B. Fayard, A. Touati, E. Sage, F. Abel, C. Champion, A. Chetoui, Single- and double-strand breaks induced in plasmid DNA irradiated by ultra-soft X-rays, J. Chim. Phys. 96 (1999) 147–151.
[48] K. Hieda, T. Hirono, A. Azami, M. Suzuki, Y. Furusawa, H. Maezawa, N. Usami, A. Yokoya, K. Kobayashi, Single- and double-strand breaks in pBR322 plasmid DNA by monochromatic X-rays on and off the K-absorption peak of phosphorus, Int. J. Radiat. Biol. 70 (1996) 437–445.
[49] A. Eschenbrenner, M.A. Herve Du Penhoat, A. Boissiere, G. Eot-Houllier, F. Abel, M.-F. Politis, A. Touati, E. Sage, A. Chetioui, Strand breaks induced in plasmid DNA by ultrasoft X-rays: influence of hydration and packing, Int. J. Radiat. Biol. 83 (2007) 687–697.
[50] P.K. Agrawala, A. Eschenbrenner, M.A. Hervé du Penhoat, A. Boissiere, M.F. Politis, A. Touati, E. Sage, A. Chetioui, Induction and repairability of DNA damage caused by ultrasoft X-rays: role of core events, Int. J. Radiat. Biol. 84 (2008) 1093–1103.
[51] K. Fujii, N. Shikazono, A. Yokoya, Nucleobase lesions and strand breaks in dry DNA thin film selectively induced by monochromatic soft X-rays, J. Phys. Chem. B 113 (2009) 16007–16015.
[52] N.J. Tao, S.M. Lindsay, A. Rupprecht, Structure of DNA hydration shells studied by Raman spectroscopy, Biopolymers 28 (1989) 1019–1030.
[53] A. Urushibara, N. Shikazono, P. O'Neill, K. Fujii, S. Wada, A. Yokoya, LET dependence of the yield of single-, double-strand breaks and base lesions in fully hydrated plasmid DNA films by 4He2+ ion irradiation, Int. J. Radiat. Biol. 84 (2008) 23–33.
[54] T. Shiina, R. Watanabe, I. Shiraishi, M. Suzuki, Y. Sugaya, K. Fujii, A. Yokoya, Induction of DNA damage, including abasic sites, in plasmid DNA by carbon ion and X-ray irradiation, Radiat. Environ. Biophys. 52 (2013) 99–112.
[55] T. Kai, A. Yokoya, M. Ukai, K. Fujii, M. Higuchi, R. Watanabe, Dynamics of low-energy electrons in liquid water with consideration of coulomb interaction with positively charged water molecules induced by electron collision, Radiat. Phys. Chem. 102 (2014) 16–22.
[56] T. Kai, A. Yokoya, M. Ukai, R. Watanabe, Cross sections, stopping powers, and energy loss rates for rotational and phonon excitation processes in liquid water by electron impact, Radiat. Phys. Chem. 108 (2015) 13–17.
[57] T. Kai, A. Yokoya, M. Ukai, K. Fujii, R. Watanabe, Thermal equilibrium and prehydration processes of electrons injected into liquid water calculated by dynamic Monte Carlo method, Radiat. Phys. Chem. 115 (2015) 1–5.
[58] T. Kai, A. Yokoya, M. Ukai, K. Fujii, R. Watanabe, Dynamic behavior of secondary electrons in liquid water at the earliest stage upon irradiation: implications for DNA damage localization mechanism, J. Phys. Chem. A 120 (2016) 8228–8233.
[59] T. Kai, A. Yokoya, M. Ukai, K. Fujii, T. Toigawa, R. Watanabeb, A significant role of non-thermal equilibrated electrons in the formation of deleterious complex DNA damage, Phys. Chem. Chem. Phys. 20 (2018) 2838–2844.
[60] K. Akamatsu, N. Shikazono, T. Saito, Fluorescence anisotropy study of radiation-induced DNA damage clustering based on FRET, Anal. Bioanal. Chem. 413 (2021) 1185–1192.
[61] T. Nakano, K. Akamatsu, M. Tsuda, A. Tujimoto, R. Hirayama, T. Hiromoto, T. Tamada, H. Ide, N. Shikazono, Formation of clustered DNA damage in vivo upon irradiation with ionizing radiation: visualization and analysis with atomic force microscopy, Proc. Natl. Acad. Sci. U. S. A. 119 (2022) e2119132119.
[62] M. Maeda, N. Usami, K. Kobayashi, Low-dose hypersensitivity in nucleus-irradiated V79 cells studied with synchrotron X-ray microbeam, J. Radiat. Res. 49 (2008) 171–180.

[63] H. Zhou, M. Hong, Y. Chai, T.K. Hei, Consequences of cytoplasmic irradiation: studies from microbeam, J. Radiat. Res. 50 (2009) A59–A65.
[64] J. Thacker, The use of recombinant-DNA techniques to study radiation-induced damage, repair and genetic change in mammalian-cells, Int. J. Radiat. Biol. 50 (1986) 1–30.
[65] A.J. Rainbow, Repair of radiation-induced DNA breaks in human adenovirus, Radiat. Res. 60 (1974) 155–164.
[66] H. Nakaue, Y. Obata, K. Kaminaga, N. Akimitsu, A. Yokoya, Visualization of the DNA repair process in mammalian cells transfected with Egfp-expressing plasmid DNA after exposure to X-rays in vitro, Radiat. Prot. Dosimetry 183 (2019) 79–83.
[67] Y. Obata, H. Nakaue, K. Hirasaki, N. Akimitsu, A. Yokoya, Expression of an X-irradiated EGFP-expressing plasmid transfected into non-irradiated human cells, Radiat. Res. 196 (2021) 261–271.

CHAPTER SEVEN

Auger electrons and DNA double-strand breaks studied by using iodine-containing chemicals

Yuya Higashi[a,†], Yue Ma[a,b,†], Kotaro Matsumoto[a], Ayumi Shiro[c], Hiroyuki Saitoh[c], Tetsuya Kawachi[d], and Fuyuhiko Tamanoi[a,*]

[a]Institute for Integrated Cell-Material Sciences, Institute for Advanced Study, Kyoto University, Kyoto, Japan
[b]Institute of Biomedical Engineering, Chinese Academy of Medical Sciences and Peking Union Medical College, Tianjin, China
[c]Kansai Photon Science Institute, Quantum Beam Science Research Directorate, National Institutes for Quantum Science and Technology, Hyogo, Japan
[d]Kansai Photon Science Institute, Quantum Beam Science Research Directorate, National Institute for Quantum Science and Technology, Kizu, Japan
*Corresponding author:

Contents

Abstract

Irradiation of high Z elements such as iodine, gold, gadolinium with monochromatic X-rays causes photoelectric effects that include the release of Auger electrons. Decay of radioactive iodine such as I-123 and I-125 also results in multiple events and some involve the generation of Auger electrons. These electrons have low energy and travel only a short distance but have a strong effect on DNA damage including the generation of double-strand breaks. In this chapter, we focus on iodine and discuss various studies

[†] The first two authors contributed equally to the preparation of this manuscript.

The Enzymes, Volume 51
ISSN 1874-6047
https://doi.org/10.1016/bs.enz.2022.08.007

that used iodine-containing chemicals to generate Auger electrons and cause DNA double-strand breaks. First, DNA synthesis precursors containing iodine were used to place iodine on DNA. DNA binding dyes such as iodine Hoechst were investigated for Auger electron generation and DNA breaks. More recently, iodine containing nanoparticles were developed. We describe our study using tumor spheroids loaded with iodine nanoparticles and synchrotron-generated monochromatic X-rays. This study led to the demonstration that an optimum effect on DNA double-strand break formation is observed with a 33.2keV X-ray which is just above the K-edge energy of iodine.

1. Introduction

X-ray irradiation of high Z elements such as gadolinium, iodine, gold and silver results in the release of Auger electrons. The Auger effect described by Pierre Auger and Lise Meitner (reviewed in Refs. [1–3]) postulates a process that involves electron release starting with the absorption of the X-ray energy by K-shell electrons (Fig. 1). This results in destabilization of the atom which is corrected by having an electron from an outer shell drop to the K-shell. This releases energy which is then absorbed by another electron leading to the release of more electrons from the atom. Thus, a

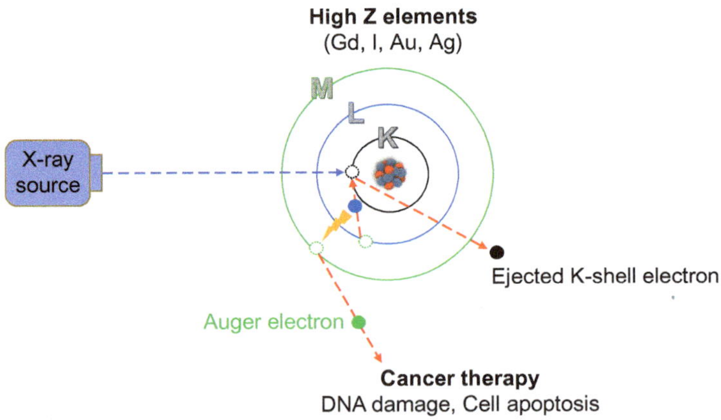

Fig. 1 The Auger effect by X-ray irradiation to high Z element. Irradiation of high Z element such as Gd, I, Au or Ag with a monochromatic X-ray having a defined energy level causes effects that include ejection of a K-shell electron (black filled circle). One of the ways to reestablish energy equilibrium is the movement of an electron from a higher shell (blue filled circle) causing the release of energy that is used by another electron (green filled circle) which is released as an Auger electron. The Auger electron has strong cell killing effect. *Reproduced from F. Tamanoi, K. Matsumoto, T.L.H. Doan, A. Shiro, H. Saitoh, Studies on the exposure of gadolinium containing nanoparticles with monochromatic X-rays drive advances in radiation therapy, Nanomaterials, 10 (2020) 1341.*

cascade of events leads to the release of multiple electrons occurs. Decay of radionucleotides such as I-123 and I-125 also causes events that include the release of the Auger electrons. The released Auger electrons have low energy and a short travel distance, but have the ability to cause single-strand and double-strand breaks (DSBs) in DNA, resulting in cell killing [4–7].

In this chapter, we focus on iodine-containing chemicals. Iodine is a more economical high Z element than gold or gadolinium. In addition, iodine has a K-edge energy of 33.2 keV which is much lower than that of gadolinium or gold. Monochromatic X-rays having energy lower than 50 keV may be generated by the use of a compact X-ray source [8]. Thus, iodine-based chemicals may be used extensively in hospital settings in the future. In fact, iodine has already been used in hospitals as an enhancing agent for CT scan [9,10]. Various chemicals that contain iodine have been developed and studied for their effect on DNA, cancer cells and tumor mass. The iodine containing chemicals developed include nucleotides such as iododeoxyuridine which can be used to incorporate iodine into DNA. We will also discuss DNA binding dyes such as iodine-Hoechst. More recently, nanoparticles containing iodine have been developed. These nanoparticles have the advantage to enable accumulation of iodine in the tumor and facilitate perinuclear localization of iodine.

Plasmid DNA, phage DNA as well as cellular DNA have been used to examine effects of Auger electrons in biological assays. With respect to cellular DNA, most studies use cultured cancer cells as a two-dimensional representation of cancer cells. We as well as others have used a tumor spheroid-based approach to study effects of Auger electrons on DNA inside cancer cells [11,12]. Tumor spheroids are three-dimensional representation of cancer cells and contain approximately 10^5 cancer cells. They mimic tumor mass and provide a convenient system for testing effects of Auger electrons.

In this review, we also describe the assay we developed that combined iodine nanoparticle-loaded tumor spheroids with synchrotron-generated monochromatic X-rays. These monochromatic X-rays have a narrow energy peak with a large number of photons (bright X-ray). This enabled us to scan the energy level of monochromatic X-ray to identify the one with optimum effect on DNA double-strand breaks [11,12].

2. Iododeoxyuridine

^{123}I and ^{125}I are radionuclides that have been conjugated to deoxyuridine (5-iodo-2′-deoxyuridine) (IUdR). This enables the incorporation of radioactive iodine into DNA, as IUdR is used as a nucleotide precursor for DNA synthesis. A variety of studies have been conducted by using cellular DNA and phage DNA that has radioactive iodine incorporated [13–17]. These studies, together with studies using DNA binding dyes, led to the following conclusions [2]. First, a single double-strand break is generated in prokaryotic cell DNA per decay of an atom of iodine-125, whereas 2–3 double-strand breaks are produced in eukaryotic cell DNA per decay. Martin and Haseltine [17] reported that damage to both strands of double-stranded oligonucleotides occurred within 1.5–2 nm from the decaying iodine-125 atom. Second, DNA double-strand breaks in mammalian cells are caused by direct as well as by indirect effects (involving reactive oxygen) on nuclear DNA. This conclusion was obtained from the analysis of the effect of the addition of DMSO (dimethylsulfoxide) which is a scavenger of reactive oxygens. Finally, it was shown that mammalian cell killing is a consequence of double-strand break formation on nuclear DNA.

Rather than using radioactive iodine, it is possible to use non-radioactive iodine. In this case, it is necessary to use external X-ray irradiation to release Auger electrons. In this method called photoactivation therapy (PAT) [18–23], iodine is incorporated into DNA by the use of IUdR and then low energy photons were used to irradiate cells. In a study by Bayart et al. [23], enhanced cellular sensitivity to radiation was observed and this was correlated with the production of DNA damage.

Use of iodinated chemicals for Auger therapy is an attractive approach for the treatment of cancer based on targeted therapy [22–26]. Radioactive iodine such as ^{125}I-IUdR as well as PAT-based assays have been investigated. In this therapy method, iodine can be placed close to or on the cellular DNA which could take advantage of the beneficial features of short travel distance and high linear energy transfer characteristics of Auger electrons. Some studies have shown promising results for treatment of cancer with Auger electron emitters, achieving tumor remissions or improved survival in patients [24]. Radioactive iodine has also been conjugated to monoclonal antibodies that bind receptors on cancer cells. An example is ^{125}I-mAb which targets cell surface receptors and induce cytotoxic effect through oxidative stress-mediated effect [27]. ^{123}I-MAPi (Iodine-123 Meitner-Auger PARP1 inhibitor) binds to the DNA repair enzyme PARP1 [28] demonstrate a survival benefit in mouse models of GBM (glioblastoma multiforme).

3. Iodine containing DNA binding chemicals

Radioactive iodine can be placed on DNA by using DNA binding dyes such as Hoechst. For example, ^{123}I-labeled Hoechst 33342 were used to place ^{123}I on DNA by the virtue of minor groove binding of Hoechst dye. [29,30]. Studies using plasmid DNA incubated with ^{123}I-labeled Hoechst 33342 highlighted the role of DNA topology in double strand break production [7,30]. Other iodine radioisotopes, iodine-125 and iodine-124 were also labeled with Hoechst 33342 and Hoechst 33258, and have been studied for DNA double-strand break (DSB) using plasmid DNA [31–35]. Acridine orange has also been used as a DNA binding dye [36]. Studies have shown that the distance between the ^{125}I atom and the DNA helix is important for double-strand breaks by low-energy Auger emitter [31]. Balagurumoorthy et al. [31] synthesized a series of ^{125}I-labeled Hoechst derivatives, and the variation of ^{125}I-DNA distance with DSB yields was investigated. The results suggested that DSB production switches from direct action to indirect action at a critical distance between the ^{125}I atom and DNA helix.

4. Nanoparticles containing iodine

Nanoparticles have a number of advantageous features. They not only enable cellular uptake of high Z elements but also can achieve perinuclear localization of high Z elements which place high Z elements close to DNA. Nanoparticles are advantageous for cancer therapy, as they can be targeted to tumors. They can be accumulated in the tumor due to passive as well as active mechanisms. Indeed, *in vivo*, the high vascularization and porosity of the blood vessel wall lead to preferential accumulation of nanoparticles in the tumor cells compared to healthy cells. This phenomenon, called the Enhanced Permeability and Retention (EPR) effect has been widely used to preferentially target cancer cells. Surface modification can also be performed as an active targeting strategy [37–39]. Nanoparticles containing high Z elements such as gold, gadolinium and iodine have been developed [11,12,40–45]. These nanosized particles (less than 300 nm diameter) have a number of properties that are beneficial for the Auger therapy. As mentioned above, they are easily taken up into cancer cells and are localized at the periphery of cell nucleus, increasing the chance to damage cancer cell DNA [40]. The high Z element loaded nanoparticles were shown to enhance radiation sensitivity in cultured cells and in animal models when irradiated with monochromatic X-ray [41–45].

There are numerous reports on the development of nanoparticles containing radioactive iodine-131 [46,47]. Prihatiningsih et al. [47] developed PSN-^{131}I, porous silica nanoparticles containing radioactive iodine-131. The synthesized nanoparticles were 100–150 nm in size and were not toxic to normal cell samples up to high concentrations. The PSN-^{131}I showed a promising uptake efficiency in cancer cells.

Hainfeld et al. [48–51] developed 20 nm small iodine nanoparticles (INPs) synthesized by cross-linking triiodobenzene. INPs are coated by PEG to confer a long half-life of 40 h in blood. These nanoparticles are preferentially taken up into tumors. Intravenous injection of INPs into mice with human breast cancer in the brain, and subsequent irradiation with X-rays enhanced the effect of radiation, resulting in an extended lifespan of the diseased mice.

We have developed iodine containing nanoparticles by incorporating iodine into porous silica nanoparticles [11] (Fig. 2A). The method combines

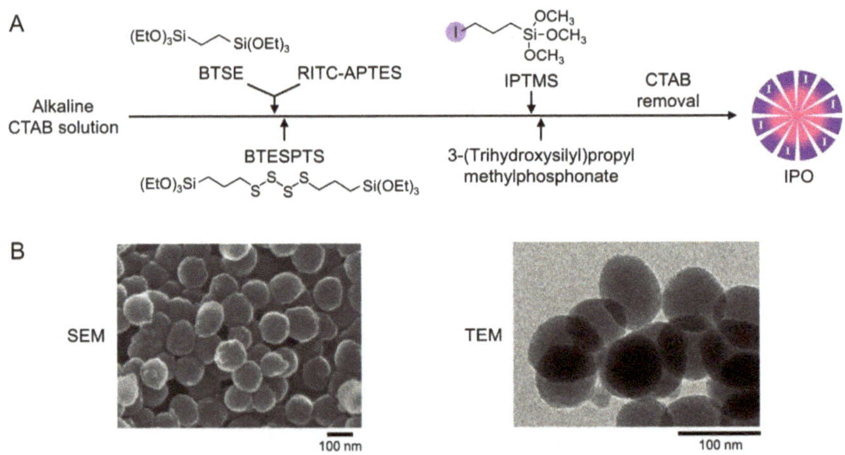

Fig. 2 Iodine containing porous organosilica (IPO) nanoparticles. (A) Synthetic scheme of IPO nanoparticles. BTESPTS, bis[3-(triethoxysilyl)propyl] tetrasulfide; BTSE, 1,2-bis(triethoxysilyl)ethane; IPTMS, (3-iodopropyl)trimethoxysilane. (B) Images of IPO nanoparticles obtained by scanning electron microscope (SEM) and transmission electron microscope (TEM). *Reproduced from Y. Higashi, K. Matsumoto, H. Saitoh, A. Shiro, Y. Ma, M. Laird, et al., Iodine containing porous organosilica nanoparticles trigger tumor spheroids destruction upon monochromatic X-ray irradiation: DNA breaks and K-edge energy X-ray, Sci. Rep. 11 (2021) 14192.*

the iodine-containing precursor with organosilica precursors that were previously used for organosilica nanoparticles [52,53]. These nanoparticles are relatively stable and possess a surface that can be modified [54,55]. Because they contain tetrasulfide bonds that are degradable under reducing conditions, these nanoparticles are biodegradable. They are named IPO for iodine-containing porous organosilica nanoparticles (IPO) [11]. The synthesis protocol of IPO involved one-pot sol-gel method of organosilica nanoparticles using organic alkoxysilane precursors, bis[3-(triethoxysilyl) propyl] tetrasulfide (BTESPTS) and 1,2-bis(triethoxysilyl)ethane (BTSE) [11]. The reaction was performed in alkaline solution at 80 °C containing cetyltrimethylammonium bromide (CTAB) which is a surfactant to promote the pore formation in nanoparticles. Aminopropyltriethoxysilane (APTES) bound with rhodamine-B isocyanate (RITC) was also incorporated into the silica network to detect the nanoparticles by fluorescence under a microscope. To incorporate iodine atoms into nanoparticles, iodine precursor ([3-iodopropyl]trimethoxysilane, IPTMS) was co-condensed with BTESPTS and BTSE. 3-(Trihydroxysilyl)propyl methylphosphonate was added to increase dispersibility in solution. Fig. 2B shows SEM and TEM characterization of IPO. These nanoparticles are relatively stable and possess surface that can be modified [11]. These nanoparticles were loaded efficiently onto tumor spheroids.

5. Our experiments using monochromatic X-rays and tumor spheroids incubated with iodine nanoparticles

In order to investigate optimum condition for the effect of the Auger electrons on DNA and cancer cells, we have developed an assay that used monochromatic X-rays combined with tumor spheroids loaded with IPO nanoparticles [11,12]. This study as well as our study using gadolinium nanoparticles led to the observation that an optimum condition for the Auger effect on cancer cells can be identified.

5.1 Experimental setup

A tumor spheroid is a spherical mass of cancer cells with a diameter of approximately 400 μm, which contains tens of thousands of cells (one estimate shows 7×10^4 cells). These three-dimensional tumor spheroids are useful as cancer models because they more closely resemble the state of actual cancer than two-dimensional cultured cells. Tumor spheroids can be formed by using a variety of human cancer cells including ovarian cancer cells, lung cancer cells, brain tumor cells and sarcoma cells. Tumor spheroids can in the future be modified by adding other components of human cancer such as fibroblasts and macrophages (these will be called tumor organoids).

To generate Auger electrons within the tumor spheroid cells, tumor spheroids were loaded with iodine nanoparticles by incubating with IPO for 24 h. We have shown that IPOs can be distributed throughout the tumor spheroids by using confocal microscopy. When a single cancer cell was analyzed, it was found that IPOs were localized at the periphery of cell nucleus.

X-rays are electromagnetic waves in the wavelength range of one picometer to 10 nanometers. Synchrotron generated white X-ray can be separated into monochromatic X-rays having a single energy level (sharp peak of energy). Monochromatic X-rays have been used for imaging studies to obtain sharper images. A synchrotron accelerates electrons to extremely high energy and then directs them to change directions by the use of magnets (bending magnets). This results in the emission of X-ray beams emitted from a bending magnet and are directed to beamlines that are placed surrounding the synchrotron ring. Monochromatic X-ray irradiation of tumor spheroids was performed at SPring-8 synchrotron BL14B1 [11,12]. The monochromatic X-ray generation system in BL14B1 is shown in the Fig. 3. White X-rays generated by the bending magnet source of BL14B1 are monochromatized by two silicon single crystals, and the beam was formed by vertical and horizontal slits. The beam is directed to tumor spheroids that are placed in a microtube containing medium and placed in a tube rack on an automated stage with three XYZ axes. To generate Auger

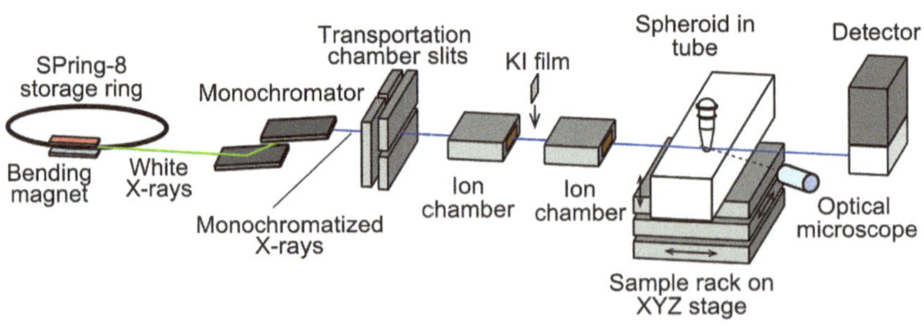

Fig. 3 Experimental setup for monochromatic X-ray irradiation to tumor spheroid. *Reproduced from Y. Higashi, K. Matsumoto, H. Saitoh, A. Shiro, Y. Ma, M. Laird, et al., Iodine containing porous organosilica nanoparticles trigger tumor spheroids destruction upon monochromatic X-ray irradiation: DNA breaks and K-edge energy X-ray, Sci. Rep. 11 (2021) 14192.*

electrons inside tumor spheroid cells, monochromatic X-rays at or higher than K-edge energy were used [11,12,40].

5.2 Detection of DNA double-strand breaks and apoptosis

Irradiation of tumor spheroids with monochromatic X-ray having 33.2 keV energy for 30 min followed by incubation for 3 days in CO_2 incubator resulted in complete destruction of tumor spheroids. The destruction was dependent on the time of irradiation. In addition, the destruction was dependent on the amount of IPO loaded onto the spheroids. Spheroids loaded with a control nanoparticle not containing iodine were not affected by the irradiation, demonstrating that the irradiation with 33.2 keV monochromatic X-ray did not affect tumor spheroids.

We investigated whether DNA double-strand breaks are induced by the irradiation by using γH2AX foci assay [11]. This method detects DNA double-strand breaks; upon the generation of double-strand breaks, histone H2AX molecules are rapidly phosphorylated (Fig. 4A). Phosphorylation of Serine-139 is one of the earliest events upon X-ray exposure and this can be detected by the use of an antibody specific for this phosphorylation. Tumor spheroids were collected immediately after the irradiation with monochromatic X-rays and processed for the γH2AX assay. As can be seen in Fig. 4B, positive signals as seen by blue fluorescence in our assay were detected with IPO loaded tumor spheroids after irradiation with 33.2 keV monochromatic X-ray. The signal was dependent on the presence of iodine, as little signal was detected with the no iodine nanoparticles, although the tumor spheroid distribution of the no iodine nanoparticles was very similar to that observed with IPO. The signal for DNA double-strand breaks was not detected after irradiation with 33.0 keV X-ray.

DNA double-strand breaks lead to apoptosis induction. This was demonstrated by the use of a TUNEL assay. This assay detects DNA breaks occurring during apoptosis which is responsible for the destruction of tumor spheroids. Apoptosis induction occurs during the incubation in CO_2 incubator for 2–3 days, as the process of apoptosis takes prolonged incubation. The apoptosis induction involves a series of biochemical reactions including activation of caspases.

5.3 Optimum energy of X-ray required for DNA double-strand breaks

Results of experiments using three different energy levels (33.0, 33.2 and 33.4 keV) demonstrated that spheroid destruction, apoptosis induction

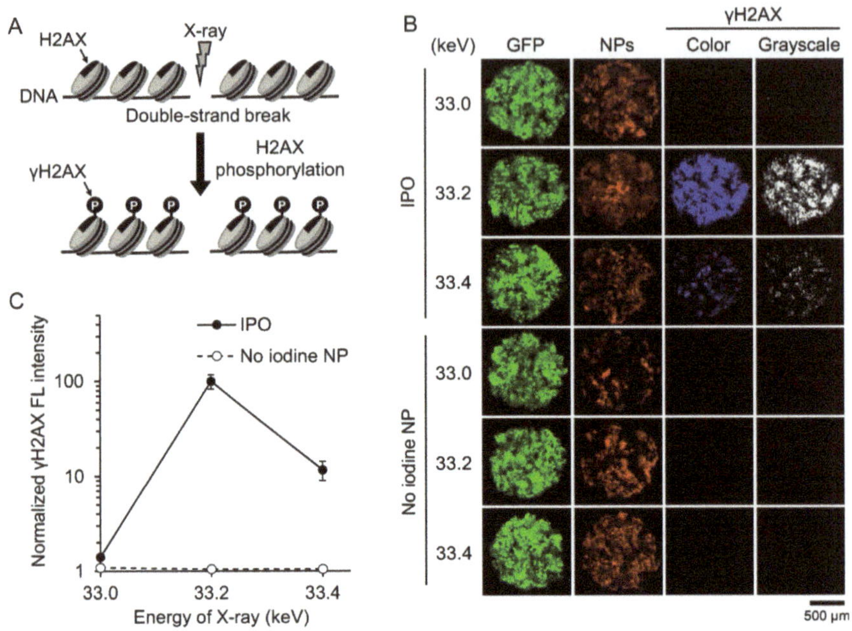

Fig. 4 Detection of DNA double-strand breaks in tumor spheroids by γH2AX assay. (A) Generation of γH2AX caused by double strand DNA breaks. (B) Microscopic images of γH2AX foci in OVCAR8 tumor spheroids incubated with IPO followed by irradiated with monochromatic X-rays. The γH2AX foci are displayed by blue (color) as well as white (grayscale). GFP, cancer cells expressing GFP; NP, nanoparticle. (C) Quantitation of double strand DNA breaks plotted against the energy of X-ray. FL, fluorescence. *Reproduced from Y. Higashi, K. Matsumoto, H. Saitoh, A. Shiro, Y. Ma, M. Laird, et al., Iodine containing porous organosilica nanoparticles trigger tumor spheroids destruction upon monochromatic X-ray irradiation: DNA breaks and K-edge energy X-ray, Sci. Rep. 11 (2021) 14192.*

and double strand DNA breaks occurred with the 33.2 keV X-ray irradiation, but not with 33.0 keV irradiation and less with 33.4 keV irradiation which suggests that an optimum energy for the DNA double-strand breaks can be identified. Interestingly, double strand DNA break signal was much weaker after 33.4 keV irradiation compared with the signal observed after 33.2 keV irradiation (Fig. 4C).

To further investigate whether an optimum energy can be identified for the DNA double-strand break formation by Auger electron related events, X-rays with 0.1 keV intervals were used. This was possible, as X-rays with very sharp peak could be generated at SPring-8. As can be seen in Fig. 5, DNA double-strand break formation was maximum when a monochromatic X-ray with 33.2 keV was used. Significant decrease was observed with

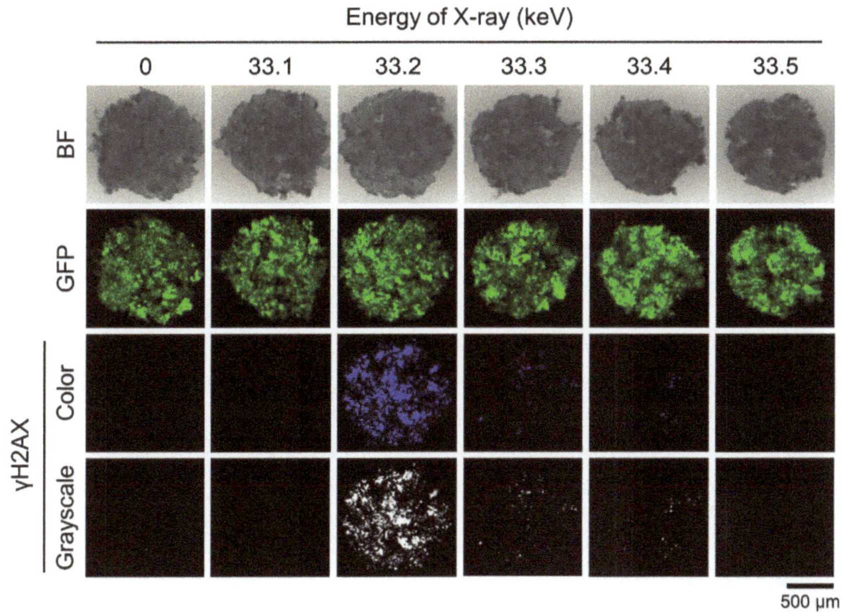

Fig. 5 Detection of DNA double-strand breaks in tumor spheroids irradiated with X-rays at 0.1 keV intervals. Microscopic images of γH2AX foci in OVCAR8 tumor spheroids loaded with IPO followed by irradiated with monochromatic X-rays are displayed. The γH2AX foci are shown by blue (color) as well as white (grayscale). BF, bright field; GFP, cancer cells expressing GFP. *Reproduced from Y. Higashi, K. Matsumoto, H. Saitoh, A. Shiro, Y. Ma, M. Laird, et al., Iodine containing porous organosilica nanoparticles trigger tumor spheroids destruction upon monochromatic X-ray irradiation: DNA breaks and K-edge energy X-ray, Sci. Rep. 11 (2021) 14192.*

33.3 keV X-ray. These results are consistent with the photoelectric effects of iodine that are induced when X-ray of K-edge energy (33.2 keV) is used, as X-ray energy is absorbed by K-shell electrons. However, the energy absorption alone does not explain the decrease in efficiency of tumor spheroid destruction with X-rays having energy levels higher than the K-edge energy. Multiple events including the release of the Auger electrons happen after the K-edge energy absorption. The energy of electrons released may differ depending on the energy absorbed. It will be interesting to follow the Auger emission after irradiating with different energy X-rays.

6. Experiments with gadolinium-loaded nanoparticles

Our finding shows that iodine nanoparticles (IPO) can be used to induce DNA double-strand breaks by irradiating with a monochromatic

X-ray. We further showed that an optimum effect can be achieved with a particular level of X-ray energy. In the case of iodine, this is 33.2 keV. A similar experiment using tumor spheroids loaded with gadolinium containing nanoparticles has identified 50.25 keV as an optimum energy of monochromatic X-ray that could cause near complete destruction of the tumor spheroids loaded with mesoporous silica nanoparticles containing gadolinium (Gd-MSN) [12,40]. This effect was exposure time dependent and was also dependent on the amount of gadolinium loaded to spheroids. On the other hand, little destruction occurs when exposed to X-ray of 50.0 keV. This dramatic difference between the effect of 50.25 keV and 50.0 keV X-rays is consistent with the idea that the Auger electrons are exerting cellular effect. When X-rays of 50.4 keV were used, tumor spheroid destruction also occurred, but we noted that the destruction efficiency was slightly less compared to what happens when exposed to 50.25 keV X-rays.

7. Summary

Because of its nanometer short range, Auger electrons have been thought as a potential way for targeted cancer therapy. Auger electron emitters such as iodine radionuclides and DNA binding dyes containing iodine have been developed. In addition, nanoparticles containing iodine have been developed. Recent work in our laboratory led to the development of iodine containing mesoporous silica nanoparticles called IPO. We have shown that monochromatic X-ray irradiation of tumor spheroids containing IPO results in DNA double-strand breaks and that an optimum effect can be seen with 33.2 keV monochromatic X-ray.

Acknowledgment

This work was supported by JSPS KAKENHI Grant Nos. JP20H00331 and JP21K14502.

References

[1] P. Auger, The auger effect, Surf. Sci. 48 (1) (1975) 1–8.
[2] A.I. Kassis, The amazing world of auger electrons, Int. J. Radiat. Biol. 80 (11 – 12) (2004) 789–803.
[3] J. Kawai, Theory of radiative Auger effect—an alternative to X-ray absorption spectroscopy, J. Electron Spectrosc. Relat. Phenom. 101–103 (1999) 847–850.
[4] T. Goorley, R. Zamenhof, H. Nikjoo, Calculated DNA damage from gadolinium Auger electrons and relation to dose distributions in a head phantom, Int. J. Radiat. Biol. 80 (11–12) (2004) 933–940.
[5] R.F. Martin, L.E. Feinendegen, The quest to exploit the Auger effect in cancer radiotherapy—a reflective review, Int. J. Radiat. Biol. 92 (11) (2016) 617–632.

[6] A. Ku, V.J. Facca, Z. Cai, R.M. Reilly, Auger electrons for cancer therapy—a review, EJNMMI Radiopharm. Chem. 4 (1) (2019) 1–36.
[7] P. Balagurumoorthy, K. Chen, S.J. Adelstein, A.I. Kassis, Auger electron-induced double-strand breaks depend on DNA topology, Radiat. Res. 170 (1) (2008) 70–82.
[8] E. Eggl, S. Schleede, M. Bech, et al., X-ray phase-contrast tomography with a compact laser-driven synchrotron source, Proc. Natl. Acad. Sci. U. S. A. 112 (18) (2015) 5567–5572.
[9] R.S. Mello, H. Callisen, J. Winter, A.R. Kagan, A. Norman, Radiation dose enhancement in tumors with iodine, Med. Phys. 10 (1) (1983) 75–78.
[10] R. Perkerson Jr., B. Erwin, B. Baumgartner, V. Phillips, W. Torres, J. Clements Jr., et al., CT densities in delayed iodine hepatic scanning, Radiology 155 (2) (1985) 445–446.
[11] Y. Higashi, K. Matsumoto, H. Saitoh, A. Shiro, Y. Ma, M. Laird, et al., Iodine containing porous organosilica nanoparticles trigger tumor spheroids destruction upon monochromatic X-ray irradiation: DNA breaks and K-edge energy X-ray, Sci. Rep. 11 (2021) 14192.
[12] K. Matsumoto, H. Saitoh, T.L.H. Doan, A. Shiro, K. Nakai, A. Komatsu, et al., Destruction of tumor mass by gadolinium-loaded nanoparticles irradiated with monochromatic X-rays: implications for the Auger therapy, Sci. Rep. 9 (2019) 13275.
[13] S.J. Adelstein, A.I. Kassis, L. Bodei, G. Mariani, Radiotoxicity of iodine-125 and other auger-electron-emitting radionuclides: background to therapy, Cancer Biother. Radiopharm. 18 (3) (2003) 301–316.
[14] R.E. Krisch, R.D. Ley, Induction of lethality and DNA breakage by the decay of iodine-125 in bacteriophage T4, Int. J. Radiat. Biol. 25 (1974) 21–30.
[15] R.E. Krisch, et al., DNA breakage, repair and lethality after 125I decay in rec+ and recA strains of Escherichia coli, Int. J. Radiat. Biol. 29 (1976) 37–50.
[16] A. Schmidt, G. Hotz, The occurrence of double-strand breaks in coliphage T1-DNA by iodine-125 decay, Int. J. Radiat. Biol. 24 (1973).
[17] R.F. Martin, W.A. Haseltine, Range of radiochemical damage to DNA with decay of iodine-125, Science 213 (1981) 896–898.
[18] R. Fairchild, A. Brill, K. Ettinger, Radiation enhancement with iodinated deoxyuridine, Investig. Radiol. 17 (4) (1982) 407–416.
[19] A. Yokoya, T. Ito, Photon-induced Auger effect in biological systems: a review, Int. J. Radiat. Biol. 93 (2017) 743–756.
[20] B.H. Laster, W.C. Thomlinson, R.G. Fairchild, Photon activation of iododeoxyuridine: biological efficacy of Auger electrons, Radiat. Res. 133 (1993).
[21] J.P. Dugas, et al., Dependence of cell survival on iododeoxyuridine concentration in 35-keV photon-activated Auger electron radiotherapy, Int. J. Radiat. Oncol. Biol. Phys. 79 (2011) 255–261.
[22] D. Alvarez, et al., Impact of IUdR on rat 9L glioma cell survival for 25–35 keV photon-activated Auger electron therapy, Radiat. Res. 182 (2014) 607–617.
[23] E. Bayart, F. Pouzoulet, L. Calmels, J. Dadoun, F. Allot, J. Plagnard, et al., Enhancement of IUdR radiosensitization by low-energy photons results from increased and persistent DNA damage, PLoS One 12 (1) (2017) e0168395.
[24] M.T. Ercan, M. Caglar, Therapeutic radiopharmaceuticals, Curr. Pharm. Des. 6 (11) (2000) 1085–1121.
[25] S. Sundell-Bergman, K. Johanson, S. Richter, P. Östbergh, Enhancement of radiation sensitivity in iododeoxyuridine labelled cells exposed to low energy x-rays, Acta Oncol. 29 (5) (1990) 623–626.
[26] C. Ceberg, B.-A. Jönsson, Y. Prezado, T. Pommer, H. Nittby, E. Englund, et al., Photon activation therapy of RG2 glioma carrying Fischer rats using stable thallium and monochromatic synchrotron radiation, Phys. Med. Biol. 57 (24) (2012) 8377.

[27] S. Paillas, R. Ladjohounlou, et al., Localized irradiation of cell membrane by Auger electrons is cytotoxic through oxidative stress-mediated nontargeted effects, Antioxid. Redox Signal. 25 (2016) 467.

[28] G. Pirovani, S.A. Jannetti, et al., Targeted brain tumor radiotherapy using an Auger emitter, Clin. Cancer Res. 26 (2020) 2871–2881.

[29] A.I. Kassis, M.A. Walicka, S.J. Adelstein, Double-strand break yield following ^{125}I decay—effects of DNA conformation, Acta Oncol. 39 (6) (2000) 721–726.

[30] P. Balagurumoorthy, K. Wang, S.J. Adelstein, A.I. Kassis, DNA double-strand breaks induced by decay of ^{123}I-labeled Hoechst 33342: role of DNA topology, Int. J. Radiat. Biol. 84 (12) (2008) 976–983.

[31] P. Balagurumoorthy, X. Xu, K. Wang, S.J. Adelstein, A.I. Kassis, Effect of distance between decaying (125)I and DNA on Auger-electron induced double-strand break yield, Int. J. Radiat. Biol. 88 (12) (2012) 998–1008.

[32] P.N. Lobachevsky, R.F. Martin, Plasmid DNA breakage by decay of DNA-associated auger emitters: experiments with 123I/125I-iodoHoechst 33258, Int. J. Radiat. Biol. 80 (11–12) (2004) 915–920.

[33] P.N. Lobachevsky, R.F. Martin, An improved approach to the analysis of plasmid DNA breakage by decay of DNA-associated auger emitters, Int. J. Radiat. Biol. 80 (11–12) (2004) 861–866.

[34] P. Balagurumoorthy, K. Chen, R.C. Bash, S.J. Adelstein, A.I. Kassis, Mechanisms underlying production of double-strand breaks in plasmid DNA after decay of 125I-Hoechst, Radiat. Res. 166 (2) (2006) 333–344.

[35] A.I. Kassis, R.S. Harapanhalli, S.J. Adelstein, Comparison of strand breaks in plasmid DNA after positional changes of Auger electron-emitting iodine-125, Radiat. Res. 151 (2) (1999) 167–176.

[36] E. Pereira, L. Do Quental, E. Palma, M.C. Oliveira, F. Mendes, P. Raposinho, et al., Evaluation of acridine orange derivatives as DNA-targeted radiopharmaceuticals for Auger therapy: influence of the radionuclide and distance to DNA, Sci. Rep. 7 (2017) 42544.

[37] S. Patel, J. Kim, M. Herrera, A. Mukherjee, A.V. Kabanov, G. Sahay, Brief update on endocytosis of nanomedicines, Adv. Drug Deliv. Rev. 144 (2019) 90–111.

[38] Y. Shi, R. Van der Meel, X. Chen, T. Lammers, The EPR effect and beyond: strategies to improve tumor targeting and cancer nanomedicine treatment efficacy, Theranostics 10 (17) (2020) 7921.

[39] J.M. Rosenholm, A. Meinander, E. Peuhu, R. Niemi, J.E. Eriksson, C. Sahlgren, et al., Targeting of porous hybrid silica nanoparticles to cancer cells, ACS Nano 3 (1) (2009) 197–206.

[40] F. Tamanoi, K. Matsumoto, T.L.H. Doan, A. Shiro, H. Saitoh, Studies on the exposure of gadolinium containing nanoparticles with monochromatic X-rays drive advances in radiation therapy, Nanomaterials 10 (7) (2020) 1341.

[41] S.-J. Seo, S.-M. Han, J.-H. Cho, K. Hyodo, A. Zaboronok, H. You, et al., Enhanced production of reactive oxygen species by gadolinium oxide nanoparticles under core–inner-shell excitation by proton or monochromatic X-ray irradiation: implication of the contribution from the interatomic de-excitation-mediated nanoradiator effect to dose enhancement, Radiat. Environ. Biophys. 54 (4) (2015) 423–431.

[42] A. Detappe, E. Thomas, M.W. Tibbitt, S. Kunjachan, O. Zavidij, N. Parnandi, et al., Ultrasmall silica-based bismuth gadolinium nanoparticles for dual magnetic resonance–computed tomography image guided radiation therapy, Nano Lett. 17 (3) (2017) 1733–1740.

[43] S. Kotb, A. Detappe, F. Lux, F. Appaix, E.L. Barbier, V.-L. Tran, et al., Gadolinium-based nanoparticles and radiation therapy for multiple brain melanoma metastases: proof of concept before phase I trial, Theranostics 6 (3) (2016) 418.

[44] F. Lux, V.L. Tran, E. Thomas, S. Dufort, F. Rossetti, M. Martini, et al., AGuIX® from bench to bedside—transfer of an ultrasmall theranostic gadolinium-based nanoparticle to clinical medicine, Br. J. Radiol. 92 (1093) (2019) 20180365.
[45] Z. Kuncic, S. Lacombe, Nanoparticle radio-enhancement: principles, progress and application to cancer treatment, Phys. Med. Biol. 63 (2) (2018) 02TR1.
[46] L. Zhang, H. Wang, Construction of radioactive iodine nanoparticle for active targeting and therapy of anaplastic thyroid cancer, J. Nucl. Med. 59 (2018) 1290.
[47] M.C. Prihatiningsih, T. Ariyanto, E.G.R. Putra, V.Y. Susilo, I. Mahendra, I. Prasetyo, Radioiodination of modified porous silica nanoparticles as a potential candidate of iodine-131 drugs vehicle, ACS Omega 7 (16) (2022) 13494–13506.
[48] J.F. Hainfeld, S.M. Ridwan, F.Y. Stanishevskiy, H.M. Smilowitz, et al., Iodine nanoparticle radiotherapy of human breast cancer growing in the brains of athymic mice, Sci. Rep. 10 (2020) 15627.
[49] S.M. Ridwan, J.F. Hainfeld, V. Ross, Y. Stanishevskiy, H.M. Smilowitz, Novel iodine nanoparticles target vascular mimicry in intracerebral triple negative human MDA-MB-231 breast tumors, Sci. Rep. 11 (2021) 1203.
[50] J.F. Hainfeld, S.M. Ridwan, Y. Stanishevskiy, H.M. Smilowitz, Iodine nanoparticles (Niodx™) for radiotherapy enhancement of glioblastoma and other cancers: an NCI nanotechnology characterization laboratory study, Pharmaceutics 14 (3) (2022) 508.
[51] J.F. Hainfeld, S.M. Ridwan, Y. Stanishevskiy, R. Panchal, D.N. Slatkin, H.M. Smilowitz, Iodine nanoparticles enhance radiotherapy of intracerebral human glioma in mice and increase efficacy of chemotherapy, Sci. Rep. 9 (2019) 4505.
[52] S. Chinnathambi, F. Tamanoi, Recent development to explore the use of biodegradable periodic mesoporous organosilica (BPMO) nanomaterials for cancer therapy, Pharmaceutics 12 (9) (2020) 890.
[53] N.X.D. Mai, A. Birault, K. Matsumoto, H.K.T. Ta, S.G. Intasa-ard, K. Morrison, et al., Biodegradable periodic mesoporous organosilica (BPMO) loaded with daunorubicin: a promising nanoparticle-based anticancer drug, ChemMedChem 15 (7) (2020) 593–599.
[54] A. Baeza, M. Colilla, M. Vallet-Regí, Advances in mesoporous silica nanoparticles for targeted stimuli-responsive drug delivery, Expert Opin. Drug Deliv. 12 (2) (2015) 319–337.
[55] M. Vallet-Regi, F. Tamanoi, Overview of studies regarding mesoporous silica nanomaterials and their biomedical application, Enzyme 43 (2018) 1–10.

CHAPTER EIGHT

Carbon ion radiation and clustered DNA double-strand breaks

Atsushi Shibata*
Gunma University Initiative for Advanced Research, GIAR, Gunma University, Maebashi, Japan
*Corresponding author: e-mail address: shibata.at@gunma-u.ac.jp

Contents

Abstract

A carbon ion categorized as a heavy ion particle has been used for cancer radiotherapy. High linear energy transfer (LET) carbon ion irradiation deposits energy at a high density along a particle track, generating multiple types of DNA damage. Complex DNA lesions, comprising DNA double-strand breaks (DSBs), single-strand breaks, and base damage within 1–2 helical turns (<3–4nm), are thought to be difficult to repair and critically influence cell viability. In addition to the effect of lesion complexity, the most recent studies have demonstrated another characteristic of high LET particle radiation-induced DNA damage, clustered DSBs. Clustered DSBs are defined as the formation of multiple DSBs in close proximity where the scale of clustering is approximately 1–2 μm^3, i.e., the scale of the event is estimated to be $> \sim 1$ Mbp. This chapter reviews the hallmarks of clustered DSBs and how such DNA damage influences genome instability and cell viability in the context of high LET carbon ion radiotherapy.

1. Introduction

Particle radiation is used as the next generation of cancer radiotherapy because it deposits high energy at the end of the particle range, which enables specific targeting of tumors owing to a sharp dose distribution known as the Bragg peak [1–3]. High linear energy transfer (LET) carbon

The Enzymes, Volume 51
ISSN 1874-6047
https://doi.org/10.1016/bs.enz.2022.08.008

ion irradiation, categorized as heavy ion radiation, exhibits a threefold higher relative biological effectiveness for killing cancer cells than photons, such as X-rays or γ-irradiation [4]. Such a strong cell killing effect has been considered to be due to high LET-dependent unique DNA damage that is difficult to repair because carbon ion irradiation causes multiple different types of DNA damage within a limited area along the track. To support the existence of such types of DNA lesions within a limited area, in vitro biological analyses and Monte Carlo simulations have suggested the formation of "complex DNA lesions," defined as DNA damage containing DNA double-strand breaks (DSBs), single-strand breaks, and base damage within 1–2 helical turns (in this chapter, this is defined as "complex DNA lesions") (Fig. 1). This is also referred to as "clustered DNA damage" in the other chapter of this special issue written by Yokoya or a "multiple damage site" proposed by other groups) [5,6]. Although complex lesions have been considered unrepairable, recent biological studies show that most DSBs induced by high LET particle irradiation are eventually repaired, although the speed of overall DNA repair is significantly delayed [4,7,8]. Therefore, the fact that complex DNA lesions cannot be repaired may not be the exclusive factor determining the strong cell killing effect after high LET particle irradiation. In contrast, misrepair is also a factor influencing cell viability after ionizing irradiation (IR) [9]. For example, if DSBs occur on two different chromosomes, interchromosomal exchange can occur,

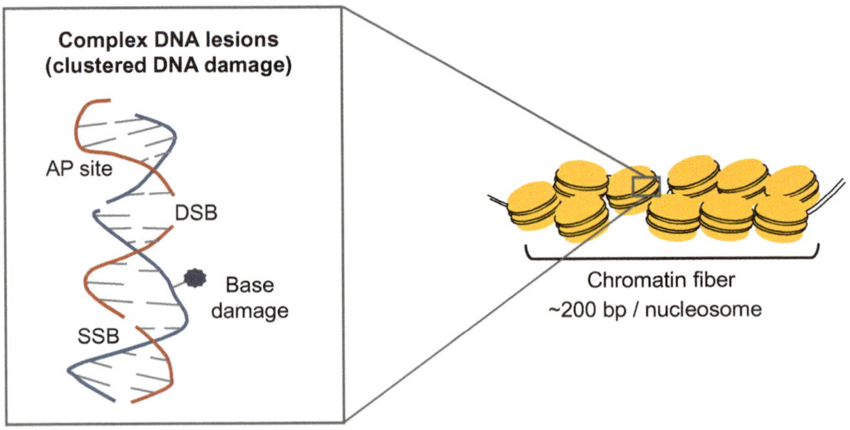

Fig. 1 Complex DNA lesions is a representative characteristic of high LET particle radiation-induced DNA damage. Complex DNA lesions (clustered DNA damage) comprising DNA double-strand breaks (DSBs), single-strand breaks, and base damage are formed within one to two helical turns after high LET particle radiation.

such as the formation of a dicentric or acentric chromosome, and the chromosome with the rearrangements fails to undergo mitosis, which results in cell death [9,10]. In addition, if a deletion within an essential gene occurs and impairs gene activity, the deletion may also be lethal. Importantly, to support the existence of chromosomal rearrangement after high LET carbon ion irradiation, the frequency of large-scale chromosomal aberrations is greater than that observed after irradiation with X-rays [11,12]. This chapter summarizes reports of clustered DSB formation along the track of high LET particle radiation and discusses how such DSBs are repaired and generate lethal mutations.

2. Clustered DSB formation visualized by high-resolution imaging

It is well known that complex DNA lesions (DNA damage through DSBs, single-strand breaks, and base damage within 1–2 helical turns) are induced by high LET particle radiation [13,14]. Details of the track structure are discussed in the other chapter of this special issue written by Yokoya et al. DSBs activate multiple DNA damage signaling pathways. A DSB triggers ataxia-telangiectasia mutated (ATM) protein kinase activation with the support of the meiotic recombination 11 (MRE11)/RAD50 DSB repair protein (RAD50)/nibrin (NBS1) [MRN] complex, which acts as a DSB sensor [15]. The activated ATM phosphorylates a thousand downstream proteins [16], including H2A histone family member X (H2AX), and promotes the recruitment of various repair proteins, including p53 binding protein 1 (53BP1), at DSB sites [15]. The phosphorylated form of H2AX is called γH2AX. Under immunofluorescence microscopy, γH2AX or 53BP1 signals are visualized as "foci" that have a round shape. Because each foci generally contains a DSB after low LET irradiation, γH2AX or 53BP1 foci (hereafter, γH2AX foci are referred to unless otherwise stated) are used as markers of DSBs [17]. However, it should be noted that since the γH2AX signals are spread across a million base pairs along the chromatin [18], even if multiple DSBs are formed within a few hundred–thousand base pairs, these breaks are visualized as a single foci, i.e., the number of γH2AX foci does not represent the number of DSBs in the case [19]. Nevertheless, generally after <3–5 Gy X-rays, clear γH2AX foci formation is identified and countable; therefore, γH2AX foci are used to quantify the number of DSBs after ionizing irradiation. As a feature of γH2AX foci after low LET radiation, γH2AX foci (DSBs) are evenly distributed in the nucleus. By contrast, when

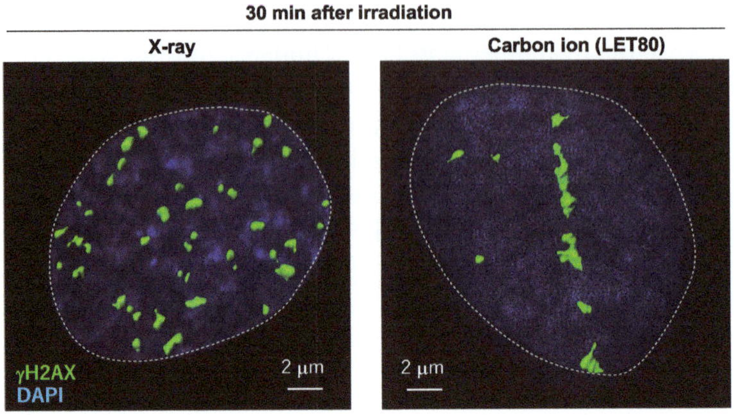

Fig. 2 Representative image of γH2AX signal obtained by conventional microscope after horizontal high LET carbon ion radiation. Polygon rendering image of γH2AX and raw image of DAPI are shown. 1BR hTERT cells were irradiated with 1 Gy carbon ion using LET 80 keV/μm or 1 Gy X-ray. Thirty minutes after ionizing radiation, cells were stained using γH2AX and DAPI.

the same dose of high LET particle irradiation is used, as it deposits high concentrated energy along the trajectory, a track of γH2AX signal, but not a focus, is observed (Fig. 2) [4]. On average, one track per nucleus is observed when 1 Gy of $>$ 60–100 keV carbon ions are horizontally irradiated [20]. By simulation, the particles deposit the energy evenly along the track [20]; however, interestingly, the γH2AX signal shows a dumpling shape along the track (Fig. 2). This may be owing to the uneven distribution of chromatin in the nucleus or the movement of chromatin [21], although the precise reason is still unknown. Under conventional microcopy, the size of γH2AX foci after high LET particle irradiation is larger than that after X-ray irradiation. Therefore, it was hypothesized that multiple DSBs were formed within the large γH2AX foci. To investigate this, advanced immunofluorescence microscopy technology was used, i.e., high-resolution imaging, a strong tool for identifying the DNA damage sites. In this chapter, three distinct scales of resolution are categorized: (1) conventional imaging, (2) deconvolution, and (3) super resolution (see also in [22]). Compared to conventional imaging, the image deconvolution method allows the visualization of multiple γH2AX micro-foci after high LET particle irradiation, although these are detected as a large lumps by conventional imaging (in this review, microfoci are defined as foci detected by deconvolution imaging) [20]. Using deconvolution imaging, $\sim$10 microfoci are detected after high LET particle irradiation (Fig. 3) [20]. Analysis of DNA repair by monitoring

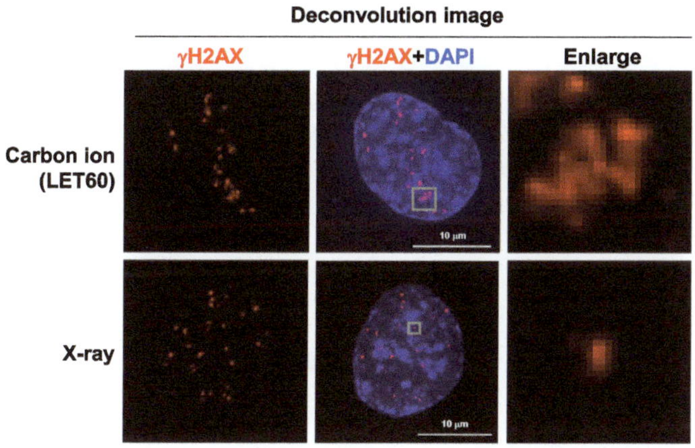

Fig. 3 Representative image of γH2AX signal after horizontal high LET carbon ion radiation using deconvolution imaging. 1BR hTERT cells were irradiated with 1 Gy carbon ion using LET 60 keV/μm. Two hours after ionizing radiation, cells were stained using γH2AX and DAPI. The image was taken by DeltaVision OMX version 4 (GE Healthcare UK Ltd.) using deconvolution imaging.

γH2AX foci disappearance at track or non-track regions showed that DSBs along the track are repaired slowly, whereas DSBs at non-track sites are repaired quickly, suggesting that DSBs formed along the track contain complex DNA lesions, which are difficult to repair [20]. In addition, this result suggests that the type of DSBs at non-track sites are likely similar to those occurring after low LET irradiation because these DSBs are repaired rapidly, and the track structure simulation shows less energy deposition at non-track regions.

More recently, the development of advanced microscopy technology allows the visualization of the recruitment of repair proteins at the single molecule level [23,24]. Three-dimensional (3D) structured illumination (3D-SIM), which is categorized as super-resolution imaging, is one of the strongest tools to visualize the cell structure in 3D using immunofluorescence. 3D-SIM imaging allows the measurement of the 3D volume of clustered γH2AX foci after carbon ion irradiation, as the resolution of the z-axis is higher than that in other types of super-resolution imaging [22]. The 3D-SIM analysis confirmed that LET carbon ion irradiation induced a larger size of γH2AX foci than that induced using X-rays. In the raw image obtained by 3D-SIM microscopy, each γH2AX signal does not represent the site of DSBs as it indicates the nucleosome nano-domain [24] (Fig. 4A); therefore, even if there is a single DSB, multiple small γH2AX foci may

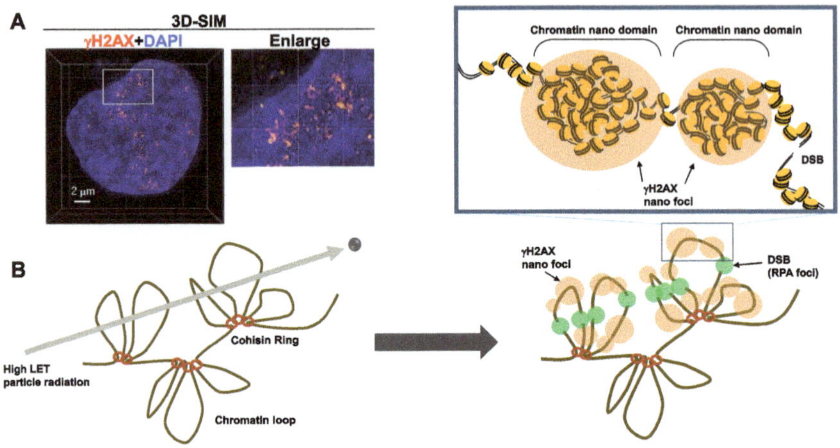

Fig. 4 Representative image of γH2AX signal after horizontal high LET carbon ion radiation using 3D-SIM setting. (A) 1BR hTERT cells were irradiated with 1 Gy carbon ion using LET 60 keV/μm. Thirty minutes after ionizing radiation, cells were stained with γH2AX and DAPI. The image was taken by DeltaVision OMX version 4 (GE Healthcare UK Ltd.) using 3D-SIM setting. (B) A schematic model showing the generation of multiple DSBs along the particle track in high order chromatin structure upon carbon ion irradiation. Under 3D-SIM image, each γH2AX nano foci do not represent a DSB. γH2AX nano foci are associated with the structure of chromatin nano domain comprising nucleosomes.

be detected using 3D-SIM. Recent studies suggest that multiple γH2AX foci are formed along the chromatin loops [25]. Thus, replication protein A (RPA), which is a marker of DSBs undergoing resection during homologous recombination (HR) [26], was monitored in our previous study [27]. Imaging of RPA and γH2AX in G2 cells shows that the large γH2AX foci encompassed multiple RPA foci after high LET carbon ion irradiation, whereas such clustered RPA foci are rarely observed after X-ray irradiation [27] (Fig. 4B). Further analysis revealed that the volume of γH2AX foci was correlated with the number of RPA foci in an LET-dependent manner, suggesting that high LET carbon irradiation induced clustered DSBs. Furthermore, the 3D analysis identified the distance between two individual RPA foci within a large γH2AX signal as being approximately 700 nm. Hence, these findings propose a model that multiple DSBs on a scale of approximately 1 μm^3 are formed along the track after high LET particle irradiation [27]. Formation of multiple numbers of DSBs in a limited area (< 100 bp) have been suggested according to the complex DNA lesion model; however, such complex lesions cannot be detected by 3D-SIM imaging. Hence, the clustered DSBs observed by super-resolution imaging is another characteristic of high LET particle radiation-induced DNA damage (note: this observation does not contradict the formation of complex DNA lesions, i.e., complex DNA lesions should be generated at each DSB site). Assuming that the diameter of nucleosomes is approximately 10 nm, 70 nucleosomes are connected in 700 nm. Assuming that one nucleosome is approximately 200 bp, there is a distance of 14,000 bp from end to end when 70 nucleosomes are lined up. However, since nucleosomes are intricately folded in the cell nucleus, if there is a spatial distance of 700 nm in the nucleus, from the viewpoint of base pairing, two DSBs may occur at intervals of 1–2 Mbp or higher, even if they are on the same chromosome. Therefore, it should be possible to detect two individual RPA foci using super-resolution imaging, even though the γH2AX signal is widely spread across the chromatin [18]. Similarly, condensed multiple RAD51 foci, which also represent the site of DSBs, are detected by stimulated emission depletion microscopy [28], which is another type of super-resolution imaging. In addition, clustered γH2AX foci have been observed when using other types of particle irradiation, such as boron and neon ions [29]. Interestingly, neon ions produce highly condensed clustered γH2AX foci and the clustering is correlated with the speed of repair, implying that levels of clustering of γH2AX foci are associated with repair difficulty due to complex DNA lesions. In addition, using single molecule imaging by transmission electron miscopy (TEM), another high-resolution imaging technology that is different from immunofluorescence imaging technology, the formation of clustered DSBs is observed along the track after high LET particle irradiation, confirming multiple DSB formation after high LET particle irradiation [30]. Using TEM technology, Lorat et al. showed that formation of multiple DSBs along the track is enriched at heterochromatic regions and that these DSBs are inefficiently repaired, suggesting that the breaks contain complex DNA lesions that are difficult to repair [31].

From the viewpoint of cancer therapy, it is important to demonstrate how the treatment kills cancer cells. In carbon ion therapy, the generation of DNA lesions that are difficult to repair is critical in demonstrating the strong cell killing effect compared to that observed after photon-based radiotherapy. To visualize the presence of clustered DSBs in a human tumor clinically treated with carbon ion radiotherapy, deconvolution based high-resolution imaging was used as the quality of samples derived from patients was not sufficient for 3D-SIM analysis [32]. Consistent with the results of tissue culture samples, clustered 53BP1 foci in a limited area are observed in a human tumor clinically treated with carbon ion radiotherapy using the spread-out Bragg peak but not in a tumor treated with X-ray radiotherapy [32]. Further analysis using a physics model showed that levels of clustered 53BP1 foci are associated with LET [32]. These data support the concept that high LET carbon ion irradiation induces clustered DSBs in a clinical setting and LET is a critical factor inducing clustered DSBs that are difficult to repair and potentially induce error prone repair, resulting in lethal mutations. In the next section, the mechanisms of repair and mutations after high LET carbon ion irradiation are introduced.

3. Repair and mutations after high LET carbon ion irradiation

In human cells, DSBs are repaired mainly by canonical non-homologous end joining (c-NHEJ) or HR after low LET irradiation, such as X-rays or γ-rays [9,10,26]. c-NHEJ functions throughout the cell cycle except during mitosis, while HR becomes active due to the increase in cyclin kinase activities during S/G2-phases; therefore, DSBs in the G1 phase are primarily repaired by c-NHEJ [26]. Importantly, the representative characteristic of DSB repair caused by high LET carbon ion irradiation is the speed of repair kinetics [8]. In G1 cells, inhibition of DNA-dependent protein kinase catalytic subunit kinase activity (DNA-PKcs) or depletion of XLF, which is a core factor of c-NHEJ, exhibits a significant repair defect even after high LET carbon ion irradiation [4,20]. By contrast, inhibition of poly(ADP-ribose) polymerase does not cause a significant repair defect [4]. This suggests that c-NHEJ mediated with DNA-PKcs is a primary repair pathway in G1 cells. Interestingly, although cells use the same c-NHEJ pathway, the kinetics of repair after high LET carbon ion irradiation is significantly slower than that after low LET irradiation [8]. This strongly suggests the presence of complex DNA lesions that are difficult to repair and requires DNA nuclease-dependent processing (see the roles of DNA nucleases for end processing in [33,34]). In G2 cells, since both c-NHEJ and HR are available [35], it is thought that cells select the optimal DSB repair pathway dependent on the type of DNA damage and DSB site [8]. After low LET irradiation, HR is used when DSBs are induced at transcriptionally active loci, which are the places that should be repaired accurately [36]. In addition, DSBs induced by high LET irradiation are preferentially repaired by HR even if DSBs are generated at non-transcriptionally active loci [8]. This is also consistent with the concept that clustered DSBs containing complex lesions should be repaired by HR because c-NHEJ may not be able to restore the loss of nucleic acids at complex DNA lesions [8]. Because HR is a slower repair pathway, the kinetics of DSB repair in G2 cells after high LET irradiation is also slower than that after X-ray irradiation [8]. Thus, the complexity of DSB ends is a factor leading to HR. Nevertheless, since pathway usage is also regulated by the dose of irradiation, transcription status, and chromatin structure around DSBs [8,31,36,37], the ratio of the use of HR, c-NHEJ, alternative-NHEJ (alt-NHEJ), or single-strand annealing is not simply explained by DSB complexity, i.e., the pathway of DSB repair must be comprehensively selected depending on the type of radiation used (see also Ref. [33]).

Some types of chromosomal translocations induced by IR are lethal [9,10]; therefore, frequency of translocation formation is associated with the efficacy of cancer cell killing after radiotherapy. The pathway generating radiation-induced transformation in G1 cells is well investigated, and c-NHEJ plays a significant role in translocation formation in human cells, although alt-NHEJ is a major pathway of translocation formation in rodent cells [34,38–40]. This difference may be explained by the distinct mechanisms of DSB end resection in different species (note: translocation formation generally requires DSB end resection) [9]. Considering the spatial distance of translocation formation, the current plausible model is defined as the contact first model [41,42]. This model suggests that mis-rejoining DSBs on two distinct chromosomes takes place preferentially when the translocation partner loci are proximal, as repair foci are moving and obtain access only if two or more foci fuse upon approach (this may be related to a repair center) [43]. Thus, the proximity of the two DSBs is a critical factor for translocation formation. In addition, the probability of foci fusion may further increase because the repair center is moving [44]. Furthermore, as 53BP1 may undergo droplet-dependent fusion [45], two repair foci may become frequently merged. Importantly, the probability of foci fusion increases if unrepaired DSBs are persistent due to the slow repair after high LET particle radiation. In human cells, chromosomes occupy territories in the interphase nucleus [46]. Therefore, if clustered DSB formation along the track overlaps at the chromosome boundary, the probability of translocation between two distinct chromosomes is increased. In our previous study, the frequency of γH2AX foci at the boundary of chromosome 1 following carbon ion irradiation is greater than fourfold higher than that after X-ray irradiation [47]. This observation confirms the notion that DSBs at the boundaries of two or more chromosomes along the track are generated after

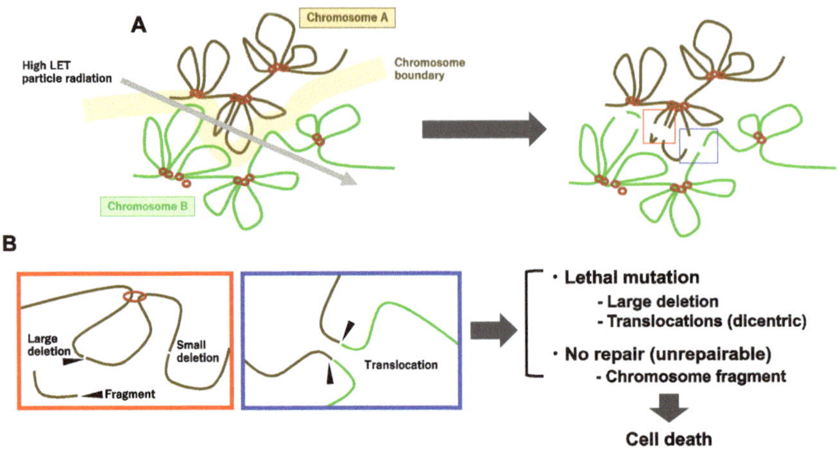

Fig. 5 Model for cell death after high LET particle irradiation. (A) Clustered DSBs are generated along the particle track after high LET particle irradiation. (B) Some DSBs can be rejoined using the original DSB end pair, whereas DSB ends that contain complex DNA lesions are slowly repaired and likely undergo error-prone repair, causing small deletions. When two DSBs generated at different region are rejoined, larger deletion is generated. If two or more DSBs are generated at the chromosome boundary and when two DSBs on different chromosomes are rejoined, chromosomal translocation is induced. As the large deletion (if an essential gene is included) or translocation (dicentric) is a lethal mutation, cells harboring such mutations undergo cell death. If a DSB that contains unrepairable complex DNA lesion is induced, such unrepairable DSBs cause chromosomal fragmentation. The essential genes located on such fragments, can render cells non-viable.

high LET particle radiation. Although only chromosome 1 was examined in that study, it is expected that these results apply to all chromosomes, as simulated by Jezkova et al. [29]. As described in the section above, because multiple DSBs in a limited area are formed both in in vitro and in clinical samples following high LET irradiation, formation of multiple DSBs within a limited area likely lead to the formation of chromosomal rearrangements, including translocations and large deletions, which become lethal mutations in cancer cells after high LET carbon ion therapy (Fig. 5).

4. Summary

X- and γ-rays induce DNA damage distributed throughout the nucleus because of low-density energy deposition. Conversely, high LET particle irradiation deposits high-density energy along the track. As previously shown, complex DNA lesions are a representative hallmark of high

LET particle ion-induced DNA damage. By contrast, advanced imaging technologies uncovered the existence of clustered DSBs within 1–2 μm in a nucleus. High LET particle radiation frequently causes dicentric chromosomes, translocations, and large deletions, including >1000 bp deletions; however, the formation of these large-scale mutations cannot be explained by a nano-scale complex lesion within 1–2 helical turns. Since the presence of clustered DSBs suggests that multiple DSBs may be formed at the boundary of two or more distinct chromosomes, these multiple DSBs within a limited area should increase the risk of translocation and large deletion formation. Because chromosomal translocations and large deletions are often lethal, this is likely an important factor in the observed high cell killing effect after high LET particle irradiation. Together, recent findings increased our understanding of how effectively high LET carbon ion irradiation kills cancer cells compared to that by conventional low LET photon therapy.

Acknowledgment

I thank Akinari Yokoya for the helpful suggestions for references about clustered DNA damage.

References

[1] T. Oike, H. Sato, S.E. Noda, T. Nakano, Translational research to improve the efficacy of carbon ion radiotherapy: experience of Gunma University, Front. Oncol. 6 (2016) 139.
[2] J.S. Loeffler, M. Durante, Charged particle therapy—optimization, challenges and future directions, Nat. Rev. Clin. Oncol. 10 (2013) 411–424.
[3] O. Mohamad, S. Yamada, M. Durante, Clinical indications for carbon ion radiotherapy, Clin. Oncol. (R. Coll. Radiol.) (2018).
[4] Y. Hagiwara, T. Oike, A. Niimi, M. Yamauchi, H. Sato, S. Limsirichaikul, K.D. Held, T. Nakano, A. Shibata, Clustered DNA double-strand break formation and the repair pathway following heavy-ion irradiation, J. Radiat. Res. 60 (2019) 69–79.
[5] D.T. Goodhead, Initial events in the cellular effects of ionizing radiations: clustered damage in DNA, Int. J. Radiat. Biol. 65 (1994) 7–17.
[6] J.F. Ward, DNA damage produced by ionizing radiation in mammalian cells: identities, mechanisms of formation, and reparability, Prog. Nucleic Acid Res. Mol. Biol. 35 (1988) 95–125.
[7] N.I. Nakajima, M. Yamauchi, S. Kakoti, L. Cuihua, R. Kato, T.B.M. Permata, M. Iijima, H. Yajima, T. Yasuhara, S. Yamada, S. Hasegawa, A. Shibata, RNF8 promotes high linear energy transfer carbon-ion-induced DNA double-stranded break repair in serum-starved human cells, DNA Repair 91–92 (2020), 102872.
[8] A. Shibata, S. Conrad, J. Birraux, V. Geuting, O. Barton, A. Ismail, A. Kakarougkas, K. Meek, G. Taucher-Scholz, M. Lobrich, P.A. Jeggo, Factors determining DNA double-strand break repair pathway choice in G2 phase, EMBO J. 30 (2011) 1079–1092.

[9] A. Shibata, P. Jeggo, A historical reflection on our understanding of radiation-induced DNA double strand break repair in somatic mammalian cells; interfacing the past with the present, Int. J. Radiat. Biol. 95 (2019) 945–956.
[10] A. Shibata, P.A. Jeggo, Canonical DNA non-homologous end-joining; capacity versus fidelity, Br. J. Radiol. 93 (2020) 20190966.
[11] K. Masumura, K. Kuniya, T. Kurobe, M. Fukuoka, F. Yatagai, T. Nohmi, Heavy-ion-induced mutations in the gpt delta transgenic mouse: comparison of mutation spectra induced by heavy-ion, X-ray, and gamma-ray radiation, Environ. Mol. Mutagen. 40 (2002) 207–215.
[12] F. Yatagai, T. Kurobe, T. Nohmi, K. Masumura, T. Tsukada, H. Yamaguchi, K. Kasai-Eguchi, N. Fukunishi, Heavy-ion-induced mutations in the gpt delta transgenic mouse: effect of p53 gene knockout, Environ. Mol. Mutagen. 40 (2002) 216–225.
[13] H. Nikjoo, P. O'Neill, W.E. Wilson, D.T. Goodhead, Computational approach for determining the spectrum of DNA damage induced by ionizing radiation, Radiat. Res. 156 (2001) 577–583.
[14] A. Ottolenghi, M. Merzagora, H.G. Paretzke, DNA complex lesions induced by protons and alpha-particles: track structure characteristics determining linear energy transfer and particle type dependence, Radiat. Environ. Biophys. 36 (1997) 97–103.
[15] A. Shibata, P.A. Jeggo, DNA double-strand break repair in a cellular context, Clin. Oncol. (R. Coll. Radiol.) (2014).
[16] S. Matsuoka, B.A. Ballif, A. Smogorzewska, E.R. McDonald 3rd, K.E. Hurov, J. Luo, C.E. Bakalarski, Z. Zhao, N. Solimini, Y. Lerenthal, Y. Shiloh, S.P. Gygi, S.J. Elledge, ATM and ATR substrate analysis reveals extensive protein networks responsive to DNA damage, Science 316 (2007) 1160–1166.
[17] M. Lobrich, A. Shibata, A. Beucher, A. Fisher, M. Ensminger, A.A. Goodarzi, O. Barton, P.A. Jeggo, gammaH2AX foci analysis for monitoring DNA double-strand break repair: strengths, limitations and optimization, Cell Cycle 9 (2010) 662–669.
[18] T. Clouaire, V. Rocher, A. Lashgari, C. Arnould, M. Aguirrebengoa, A. Biernacka, M. Skrzypczak, F. Aymard, B. Fongang, N. Dojer, J.S. Iacovoni, M. Rowicka, K. Ginalski, J. Cote, G. Legube, Comprehensive mapping of histone modifications at DNA double-Strand breaks deciphers repair pathway chromatin signatures, Mol. Cell 72 (2018) 250–262.
[19] G. Iliakis, V. Mladenova, M. Sharif, S. Chaudhary, I.V. Mavragani, A. Soni, J. Saha, A. Schipler, E. Mladenov, Defined biological models of high-let radiation lesions, Radiat. Prot. Dosimetry 183 (2019) 60–68.
[20] N.I. Nakajima, H. Brunton, R. Watanabe, A. Shrikhande, R. Hirayama, N. Matsufuji, A. Fujimori, T. Murakami, R. Okayasu, P. Jeggo, A. Shibata, Visualisation of gammaH2AX foci caused by heavy ion particle traversal; distinction between core track versus non-track damage, PLoS One 8 (2013) e70107.
[21] B. Jakob, J. Splinter, S. Conrad, K.O. Voss, D. Zink, M. Durante, M. Lobrich, G. Taucher-Scholz, DNA double-strand breaks in heterochromatin elicit fast repair protein recruitment, histone H2AX phosphorylation and relocation to euchromatin, Nucleic Acids Res. 39 (2011) 6489–6499.
[22] L. Schermelleh, R. Heintzmann, H. Leonhardt, A guide to super-resolution fluorescence microscopy, J. Cell Biol. 190 (2010) 165–175.
[23] S. Britton, J. Coates, S.P. Jackson, A new method for high-resolution imaging of Ku foci to decipher mechanisms of DNA double-strand break repair, J. Cell Biol. 202 (2013) 579–595.
[24] F. Ochs, G. Karemore, E. Miron, J. Brown, H. Sedlackova, M.B. Rask, M. Lampe, V. Buckle, L. Schermelleh, J. Lukas, C. Lukas, Stabilization of chromatin topology safeguards genome integrity, Nature 574 (2019) 571–574.

[25] C. Arnould, V. Rocher, A.L. Finoux, T. Clouaire, K. Li, F. Zhou, P. Caron, P.E. Mangeot, E.P. Ricci, R. Mourad, J.E. Haber, D. Noordermeer, G. Legube, Loop extrusion as a mechanism for formation of DNA damage repair foci, Nature 590 (2021) 660–665.
[26] A. Shibata, Regulation of repair pathway choice at two-ended DNA double-strand breaks, Mutat. Res. 803-805 (2017) 51–55.
[27] Y. Hagiwara, A. Niimi, M. Isono, M. Yamauchi, T. Yasuhara, S. Limsirichaikul, T. Oike, H. Sato, K.D. Held, T. Nakano, A. Shibata, 3D-structured illumination microscopy reveals clustered DNA double-strand break formation in widespread gammaH2AX foci after high LET heavy-ion particle radiation, Oncotarget 8 (2017) 109370–109381.
[28] J. Reindl, G.A. Drexler, S. Girst, C. Greubel, C. Siebenwirth, S.E. Drexler, G. Dollinger, A.A. Friedl, Nanoscopic exclusion between Rad51 and 53BP1 after ion irradiation in human HeLa cells, Phys. Biol. 12 (2015) 066005.
[29] L. Jezkova, M. Zadneprianetc, E. Kulikova, E. Smirnova, T. Bulanova, D. Depes, I. Falkova, A. Boreyko, E. Krasavin, M. Davidkova, S. Kozubek, O. Valentova, M. Falk, Particles with similar LET values generate DNA breaks of different complexity and reparability: a high-resolution microscopy analysis of gammaH2AX/53BP1 foci, Nanoscale 10 (2018) 1162–1179.
[30] S. Timm, Y. Lorat, B. Jakob, G. Taucher-Scholz, C.E. Rube, Clustered DNA damage concentrated in particle trajectories causes persistent large-scale rearrangements in chromatin architecture, Radiother. Oncol. 129 (2018) 600–610.
[31] Y. Lorat, S. Timm, B. Jakob, G. Taucher-Scholz, C.E. Rube, Clustered double-strand breaks in heterochromatin perturb DNA repair after high linear energy transfer irradiation, Radiother. Oncol. 121 (2016) 154–161.
[32] T. Oike, A. Niimi, N. Okonogi, K. Murata, A. Matsumura, S.E. Noda, D. Kobayashi, M. Iwanaga, K. Tsuchida, T. Kanai, T. Ohno, A. Shibata, T. Nakano, Visualization of complex DNA double-strand breaks in a tumor treated with carbon ion radiotherapy, Sci. Rep. 6 (2016) 22275.
[33] V. Mladenova, E. Mladenov, M. Stuschke, G. Iliakis, DNA damage clustering after ionizing radiation and consequences in the processing of chromatin breaks, Molecules 27 (2022).
[34] R. Biehs, M. Steinlage, O. Barton, S. Juhasz, J. Kunzel, J. Spies, A. Shibata, P.A. Jeggo, M. Lobrich, DNA double-strand break resection occurs during non-homologous end joining in G1 but is distinct from resection during homologous recombination, Mol. Cell 65 (2017) 671–684.
[35] A. Beucher, J. Birraux, L. Tchouandong, O. Barton, A. Shibata, S. Conrad, A.A. Goodarzi, A. Krempler, P.A. Jeggo, M. Lobrich, ATM and Artemis promote homologous recombination of radiation-induced DNA double-strand breaks in G2, EMBO J. 28 (2009) 3413–3427.
[36] T. Yasuhara, R. Kato, Y. Hagiwara, B. Shiotani, M. Yamauchi, S. Nakada, A. Shibata, K. Miyagawa, Human Rad52 promotes XPG-mediated R-loop processing to initiate transcription-associated homologous recombination repair, Cell 175 (2018) 558–570. e511.
[37] F. Ochs, K. Somyajit, M. Altmeyer, M.B. Rask, J. Lukas, C. Lukas, 53BP1 fosters fidelity of homology-directed DNA repair, Nat. Struct. Mol. Biol. 23 (2016) 714–721.
[38] D. Simsek, M. Jasin, Alternative end-joining is suppressed by the canonical NHEJ component Xrcc4-ligase IV during chromosomal translocation formation, Nat. Struct. Mol. Biol. 17 (2010) 410–416.
[39] H. Ghezraoui, M. Piganeau, B. Renouf, J.B. Renaud, A. Sallmyr, B. Ruis, S. Oh, A.E. Tomkinson, E.A. Hendrickson, C. Giovannangeli, M. Jasin, E. Brunet, Chromosomal translocations in human cells are generated by canonical nonhomologous end-joining, Mol. Cell 55 (2014) 829–842.

[40] R. Ceccaldi, B. Rondinelli, A.D. D'Andrea, Repair pathway choices and consequences at the double-Strand break, Trends Cell Biol. 26 (2016) 52–64.
[41] V. Roukos, T. Misteli, The biogenesis of chromosome translocations, Nat. Cell Biol. 16 (2014) 293–300.
[42] V. Roukos, T.C. Voss, C.K. Schmidt, S. Lee, D. Wangsa, T. Misteli, Spatial dynamics of chromosome translocations in living cells, Science 341 (2013) 660–664.
[43] T. Neumaier, J. Swenson, C. Pham, A. Polyzos, A.T. Lo, P. Yang, J. Dyball, A. Asaithamby, D.J. Chen, M.J. Bissell, S. Thalhammer, S.V. Costes, Evidence for formation of DNA repair centers and dose-response nonlinearity in human cells, Proc. Natl. Acad. Sci. U. S. A. 109 (2012) 443–448.
[44] M. Yamauchi, A. Shibata, K. Suzuki, M. Suzuki, A. Niimi, H. Kondo, M. Miura, M. Hirakawa, K. Tsujita, S. Yamashita, N. Matsuda, Regulation of pairing between broken DNA-containing chromatin regions by Ku80, DNA-PKcs, ATM, and 53BP1, Sci. Rep. 7 (2017) 41812.
[45] S. Kilic, A. Lezaja, M. Gatti, E. Bianco, J. Michelena, R. Imhof, M. Altmeyer, Phase separation of 53BP1 determines liquid-like behavior of DNA repair compartments, EMBO J. 38 (2019), e101379.
[46] T. Cremer, C. Cremer, Chromosome territories, nuclear architecture and gene regulation in mammalian cells, Nat. Rev. Genet. 2 (2001) 292–301.
[47] A. Niimi, M. Yamauchi, S. Limsirichaikul, R. Sekine, T. Oike, H. Sato, K. Suzuki, K.D. Held, T. Nakano, A. Shibata, Identification of DNA double strand breaks at chromosome boundaries along the track of particle irradiation, Genes Chromosom. Cancer 55 (2016) 650–660.

CHAPTER NINE

Damages of DNA in tritiated water

Yuji Hatano[a,*], Hiroaki Nakamura[b,c], Susumu Fujiwara[d], Seiki Saito[e], and Takahiro Kenmotsu[f]

[a]Hydrogen Isotope Research Center, University of Toyama, Toyama, Japan
[b]National Institute for Fusion Science, Toki, Japan
[c]Graduate School of Engineering, Nagoya University, Nagoya, Japan
[d]Faculty of Materials Science and Engineering, Kyoto Institute of Technology, Kyoto, Japan
[e]Graduate School of Science and Engineering, Yamagata University, Yamagata, Japan
[f]Faculty of Life and Medical Sciences, Doshisha University, Kyoto, Japan
*Corresponding author: e-mail address: hatano@ctg.u-toyama.ac.jp

Contents

Abstracts

Tritium is a radioisotope of hydrogen emitting low energy β-rays in disintegration to ^{3}He. DNA molecules are damaged mainly by β-ray irradiation, and additional damages can be induced by break of chemical bond by nuclear transmutation to inert ^{3}He. Deep knowledges of the mechanisms underlying DNA damages lead to better understanding of biological effects of tritium. This chapter reviews recent experimental and computer simulation activities on quantitative evaluation of damage rates by β-ray irradiation and nuclear transmutation. The rate of DNA double-strand breaks in tritiated water has been examined using a single molecule observation method. The effects of β-ray irradiation were not noticeable at the level of tritium concentration of ~kBq/cm^3, while the irradiation effects were clear at tritium concentrations of ~MBq/cm^3. The factors affecting on the DSB rate are discussed. A new image processing method for the automatic measurement of DNA length using OpenCV and deep learning is also introduced. The effects of tritium transmutation on hydrogen bonds acting between the two main

The Enzymes, Volume 51
ISSN 1874-6047
https://doi.org/10.1016/bs.enz.2022.08.009

strands of DNA have been examined using molecular dynamics simulations. The study showed that the collapsing of DNA structure by the transmutation can be quantitatively evaluated using the root mean square deviation of atomic positions.

1. Introduction

Tritium (^3H or T) is one of the major isotopes released from existing nuclear facilities and fuel of future fusion reactors. Hence, deep understanding of biological effects of tritium is important for public acceptance of those facilities. This radioisotope of hydrogen disintegrates at average half-life of 12.3 years to give ^3He and emits low energy β-rays:

$$^3_1T \rightarrow ^3_2He + \beta + \overline{\nu} + 18.6\,keV,$$

where $\overline{\nu}$ is antineutrino. The mean and maximum energy of β-rays are 5.7 and 18.6 keV, and the travel distances of β-rays in water or soft tissues at these values of energy are about 500 nm and 6 μm, respectively [1]. It means that β-rays from tritium cannot penetrate the epidermis of human skin, and, according to Ref. [2], "tritium poses a risk to health as a result of internal exposure only following ingestion in drinking water or food, or inhalation or absorption through the skin." The chemical form of tritium in a soft tissue can be divided into two groups: tissue free water tritium (TFWT) and organically bound tritium (OBT). The former is HTO in tissue water. The latter is T bound to sulfur, nitrogen or oxygen atoms (exchangeable OBT) or to carbon chain (non-exchangeable OBT) of organic molecules in cells and tissues [3].

The extensive studies have been performed on biological effects of tritium. However, large uncertainty remains in quantifying the risk of low and/or protracted radiation exposures [4]. Better understanding of the mechanisms underlying DNA damage can contributes to reduction of the uncertainty. A single molecule observation method has been developed for quantitative evaluation of the rate of double-strand breaks (DSBs) of genome-size giant DNA molecules under various conditions including exposure to visible light, ultrasound and γ-rays [5–8]. This method uses a fluorescent microscope to measure the length of each DNA molecule and allows to evaluate the rate of DSBs under well-controlled conditions including higher order structure of DNA molecules, compositions and temperature of solutions, etc. These capabilities are suitable for verification of submicron-scale computer simulation of interactions between ionizing radiation and biomolecules. It has been reported by using the single molecule observation method that the number of DSBs induced by visible light increased in proportional to the square of exposure time, while that caused by γ-rays at high dose rate (several thousands Gy/h) was simply proportional to the exposure time and irradiation dose [7]. The square root dependence on exposure time implies that DSBs are induced as the result of occurrence of two independent single-strand breaks (SSBs) nearby each other on the both sides of double-strand helix (a two-step mechanism) [7]. The linear dependence of the number of DSBs on the exposure time/radiation dose can be interpreted as that the two SSBs near by each other are simultaneously induced by a single event (a one-step mechanism) [7].

Recently, the kinetics of DSB of DNA in tritiated water was examined at far smaller dose rates (4.3×10^{-6}–1.7×10^{-2} Gy/h) [9–11]. The measurement of DSB rate at a low dose rate is a challenging issue because DSBs are induced not only by ionizing radiation but also by variety of factors including shear stress unintentionally applied during DNA handling, thermal motion, UV/visible light, microorganisms contained in a solution, etc., and the contribution of ionizing radiation to total number of DSBs decreases with decreasing dose rate. Therefore, in these studies, the effects of other factors including water temperature and sterilization were examined in addition to the irradiation effects of β-rays [10,11].

In the single molecule measurement method, it is necessary to measure the lengths of several tens–several hundreds of DNA molecules [10]. A larger number of DNA molecules is better for accurate evaluation of the number of DSBs, but it is time-consuming to measure the lengths of huge number of DNA molecules manually. The development of automatic procedures to evaluate the lengths of DNA molecules in fluorescent images greatly reduces workload of researcher and improve the statistical accuracy of measurements [12].

The disintegration of OBT to inert He results in chemical bond breaking and consequently leads to additional DNA damage. However, it is difficult to distinguish DNA damage caused by the disintegration of OBT from that induced by β-ray irradiation by experiment. Therefore, significant efforts have been made to perform computer simulation of DNA damage by the OBT disintegration [13–16].

From these viewpoints, this chapter reports the present status of the following topics:

(1) DSBs in tritiated water by β-rays irradiation at low dose rates and influences of water temperature and sterilization conditions [10,11];

(2) Automatic processing method of fluorescent images to measure DNA lengths [12]; and
(3) In silico simulation of DNA damage by tritium disintegration [13–16].

2. Double-strand breaks in tritiated water by β-rays irradiation and other factors

2.1 Experimental method

In Refs. [10, 11], genome-sized DNA molecules of bacteriophage T4 GT7 (Nippon Gene Co., Japan) were used as samples. The nominal length of the DNA molecule is 166 kilo base pairs (kbp) corresponding to 57 μm. The DNA molecules were suspended in (i) non-radioactive sterilized water for medical use, (ii) sterilized tritiated water prepared by diluting biological-grade tritiated water with sterilized water for medical use, and (iii) non-sterilized engineering-grade tritiated water prepared by diluting tritiated water with deionized water. The water (i), (ii) and (iii) are hereafter denoted as control water, bio-grade T water and eng-grade T water, respectively. The concentration of T in the water was at the levels of kBq/cm^3–MBq/cm^3. The DNA molecules were suspended in these types of water at 3–25 °C for 1–35 days. After the irradiation, fluorescent dye (YOYO-1 Iodide, ThermoFischer Scientific Co., USA) was added together with 2-mercaptoethanol and Tris-HCl buffer solution. Then, a drop of the mixed water was put on a glass substrate coated with poly-L-lysine, and the DNA molecules were extended and fixed on the surface of glass substrate by spreading the water along the surface using a pipette tip. Images of DNA molecules were acquired using a fluorescence microscope. Typical DNA images are given in the next section (Fig. 4A in 9.3.2). Length of each DNA molecule was manually measured using ImageJ software [17].

The number of DSBs occurred during immersion was evaluated by assuming DSBs took place at random positions along DNA molecules [5–8]. If DSB occurs N_{DSB} times for a DNA molecule with initial length of L_0, the molecule is broken into $N_{DSB}+1$ fragments. The sum of lengths of fragments L_n should be equal to the initial length. Namely:

$$L_1 + L_2 + L_3 + \ldots + L_{N_{DSB}} + L_{N_{DSB+1}} = L_0. \tag{1}$$

The average length L_{ave} is expressed as follows;

$$L_{ave} = \frac{L_1 + L_2 + L_3 + \ldots + L_{N_{DSB}} + L_{N_{DSB+1}}}{N_{DSB} + 1} = \frac{L_0}{N_{DSB} + 1}. \tag{2}$$

Hence, N_{DSB}, is obtained as [5–8]:

$$N_{DSB} = L_0/L_{ave} - 1. \quad (3)$$

The nominal length of DNA molecules is 57 μm, as previously mentioned. However, the real value of L_0 was smaller than the nominal length due to DSBs occurred during DNA handling (pipetting, mixing, etc.) and ~39 μm. If N_{DSB} is proportional to the immersion time t in water, as expected for a one-step mechanism,

$$L_{ave} = L_0/(\alpha t + 1), \quad (4)$$

where α is a constant.

The maximum traveling distance of β-rays from T in water is about 6 μm, as mentioned above. Hence, all energy of β-rays is absorbed by water. The values of absorption dose at T concentrations of 1 kBq/cm^3 and 1 MBq/cm^3 were estimated as the product of the mean energy (5.7 keV) and the number of disintegration to be 3.3 μGy/h and 3.3 mGy/h, respectively, where Gy = J/kg. These values are smaller than the dose rate in γ-ray irradiation performed in literature [6–8] by orders of magnitude, as mentioned above.

2.2 Quantitative evaluation of DNA double strand breaks in tritiated water

Fig. 1 shows the change in DNA length with immersion time in the control water, the bio-grade tritiated water and eng-grade tritiated water at 10 °C [10,11]. The characteristic points of this figure are as the followings.

(1) DNA length decreases with time even in the control water. It means that DSBs are induced not only by radiation but also by other factors such as thermal motion, oxidation, etc.

(2) The rate of DSBs in the bio-grade T water at 1.3 kBq/cm^3 was comparable with that in the control water. This observation suggests that the effects of β-ray irradiation is negligibly small in comparison with that of other factors such as thermal motion, oxidation, etc. at the dose rate of ~μGy/h.

(3) Significant increase in DSB rate was observed at 4.0 MBq/cm^3 corresponding to 13 mGy/h.

(4) The DSB rate in eng-grade T water was larger than that in bio-grade T water at comparable T concentrations. This difference indicates that the effects of microorganisms are significant even in the tritiated water at T concentrations as high as ~MBq/cm^3.

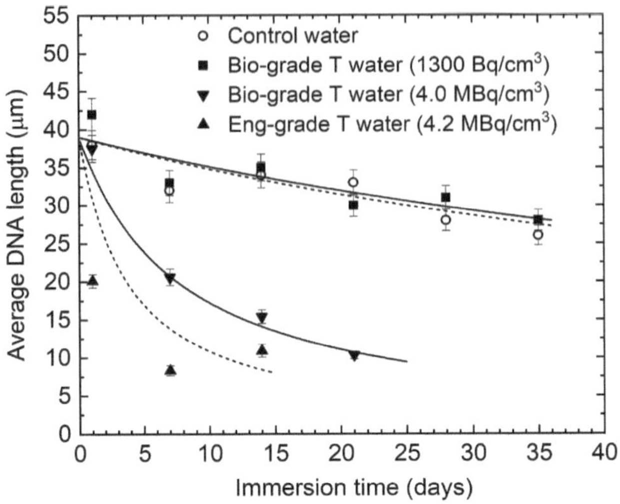

Fig. 1 Change in average DNA length with immersion time at 10°C and different dose rates (0 μGy/h in control water, 4.3 μGy/h in bio-grade T water at 1300 Bq/cm^3, 13 mGy/h in bio-grade T water at 4.0 MBq/cm^3 and 14 mGy/h in eng-grade T water at 4.2 MBq/cm^3). Lines were obtained by fitting Eq. (4) to experimental data by setting L_0=39 μm [9,10].

The change in DNA length with immersion time in the control water and the bio-grade T water was examined also at 3 °C and 25 °C in Ref. [11]. The values of N_{DSB} evaluated using Eq. (3) are plotted against immersion time for the control water and against β-ray dose for bio-grade T water. In Fig. 2B, the N_{DSB} was calculated with the difference in L_{ave} between control water and bio-grade T water to remove the influence of "pure" thermal effects. Nevertheless, it is still clear that the rates of DSBs increased with increasing temperature not only in the control water but also in the bio-grade T water. The enhancement of thermal motion and/or oxidation can be the cause of the increase in the DSB rate in the control water with increasing temperature. The temperature dependence of the DSB rate in the bio-grade T water indicates the synergy of thermal effects and irradiation effects. One of the possible mechanisms is that a SSB induced by radiation developed to DSB by thermal effects and vice versa. It is also possible that a minor damage induced by radiation developed to a strand break by thermal effects and vice versa. Fig. 3 shows the distributions of DNA lengths after irradiation with β-rays up to 4.4 Gy at 10 °C and up to 0.95 Gy at 25 °C [18]. The values of L_{ave} were 15.4 μm in the former case and 15.9 μm in

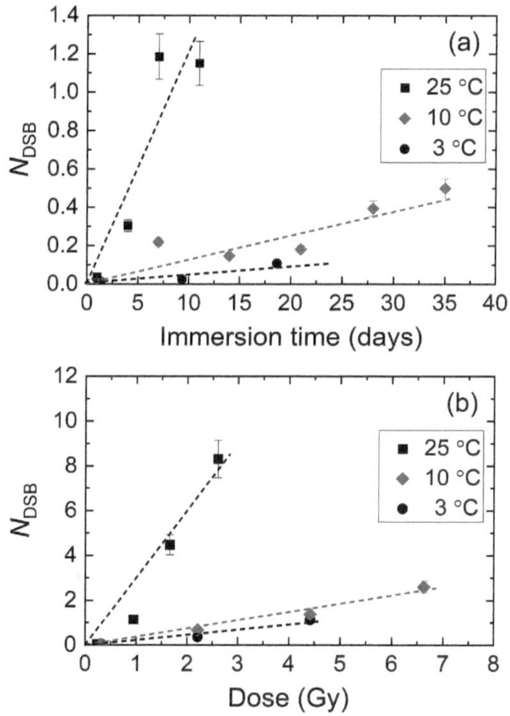

Fig. 2 Correlation between N_{DSB} and immersion time in control water (A) and that between N_{DSB} and β-ray dose in bio-grade T water (3.0–4.0 MBq/cm^3) (B) [9,10].

the latter case. In other words, the values of N_{DSB} in these two cases were comparable with each other. The distribution after irradiation at 25 °C had a sharp peak at the average DNA length, while the irradiation at 10 °C resulted in clearly broader distribution. This difference implies that the DNA molecules was broken at random positions at 10 °C but preferentially at the central part at 25 °C. Because the irradiation damage and oxidation are expected to occur at random positions, it is plausible that the stress induced by the thermal motion is the cause of development of radiation-induced SSBs to DSBs and this stress took the maximum value around the central part of DNA molecule. However, further investigation is required for more detailed understanding of the mechanisms underlying the observed synergism of thermal effects and irradiation effects. Although N_{DSB} was proportional to irradiation dose in Fig. 2B, it is inappropriate to simply consider that the DSBs were induced by the one-step mechanism.

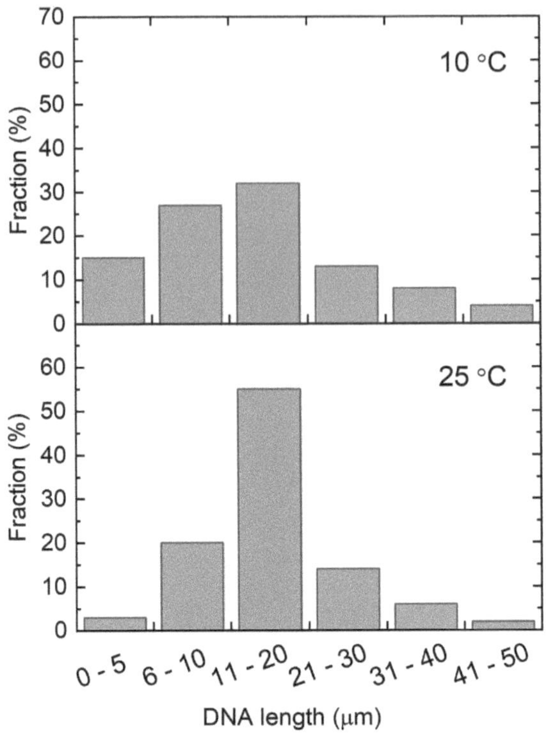

Fig. 3 The distributions of DNA lengths after irradiation with β-rays up to 4.4 Gy at 10 °C and up to 0.95 Gy at 25 °C [18]. The values of L_{ave} in the former case (15.4 μm) and in the latter case (15.9 μm) were comparable with each other.

3. Image processing method for single-molecular imaging

3.1 Background

As explained in the previous section, the single-molecule observation method is widely used for the investigation of DSBs of DNA molecules. Usually, the measurement of DNA length is done manually by experienced operators, and it is time consuming. To reduce the workload on the operators, and enable more accurate evaluations, an image processing method for automatic measurement has been developing [12]. The automation of DNA length measurement can be achieved by the following two steps: (1) extracting only DNA segments from fluorescence microscopy images, and (2) measuring the lengths of the DNA segments by OpenCV library.

Step (2) is explained in Section 3.2. Step (1) can be realized by deep learning, which is explained in Section 3.3.

3.2 Image processing by OpenCV

An example of the original image captured by a fluorescence microscope is shown in Fig. 4A. Multiple DNA segments are shown in white in the image. To measure the lengths of a DNA segments, the segments were converted into a thin line with one pixel wide (skeletonization). Using OpenCV library, the original images are skeletonized by the following steps before measuring the lengths and number of DNA segments in the image by the method explained later.

step-A1: The histogram of the image is equalized by the function "equalizeHist" in OpenCV.

step-A2: The change in local shading is reduced by the unevenness coefficient as shown in Fig. 4B.

step-A3: The histogram is manually adjusted by changing tone curve by the function "LUT" as shown in Fig. 4C.

step-A4: The image is binarized by the function "threshold" or "adaptiveThreshold" as show in Fig. 4D.

step-A5: Noise reduction is performed by the function "opening" in OpenCV as shown in Fig. 4E.

step-A6: The white pixels are skeletonized by Zhang-Suen method [19] as shown in Fig. 4F.

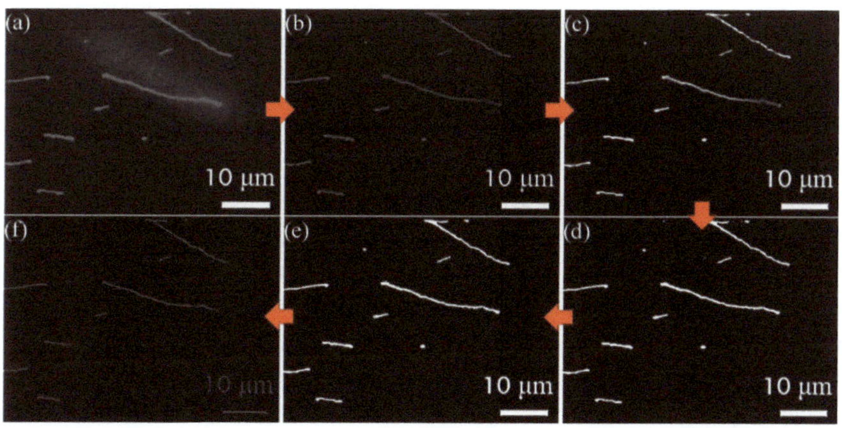

Fig. 4 Example of a result of image processing by conventional method by OpenCV [12].

Table 1 Controlled parameters used in the proposed image processing method [12]

Variables	Type	Range	Explanation
s_{uneven}	Bool	True/False	Bool value of whether to carry out step-A2. If the value is false, step-A2 is skipped
k_{uneven}	Integer	0–50	Integer value of kernel size for calculating average brightness in calculation of the unevenness coefficient. Using k_{uneven}, the value of (i,j) pixel x_{ij} is updated to x'_{ij} with the unevenness coefficient r_{ij} of (i,j) pixel by the following equations: $x_{ij}' = 255 \times (1 - r_{ij}),\ r_{ij} = (255 - x_{ij})/R_{ij}$, $R_{ij} = \frac{1}{(2k_{\text{uneven}} + 1)^2} \sum_{k,l=-k_{\text{uneven}}}^{k_{\text{uneven}}} \left(255 - x_{i+k,j+l}\right).$
s_{local}	Bool	True/False	Bool value for adaptive thresholding in step-A4. The function "adaptiveThreshold" is applied instead of "threshold" if the value is true.
$b_{\text{threshold}}$	Integer	0–255	Integer value for the threshold of the binarization in step-A4
k_{opening}	Integer	0–20	Integer value of kernel size for the function "opening"

The contrast and brightness of images depend on the experimental conditions. To support input images under various conditions, the procedure of image processing is manually controlled by the variables shown in Table 1 in addition to the tone curve in step-A3.

After the skeletonization (step-A6), the lengths and number of DNA segments in the image are measured. Fig. 5A–C show the process of measurement of the lengths of DNA segments. As shown in Fig. 5B, the skeletonized image is firstly dilated by the function "dilate" in OpenCV with kernel size of 2×2. Then, as shown in red color in Fig. 5C, the contour of the area of white pixels is detected by the function "findContours" in OpenCV. Finally, length of contour is calculated by the function "arcLength" in OpenCV. The lengths of DNA segments is obtained by dividing the length of the contour by 2. Fig. 5D and E explain the method of counting the number of DNA segments. The number can be counted by dividing the number of endpoints of the skeletonized segments by 2. It should be noted that this measurement method provides the lengths of all intersecting segments as one segment. However, average lengths of DNA segments which is used for the calculation of number of DSBs in DNA is obtained by dividing the total lengths of all DNA segments by the number of segments.

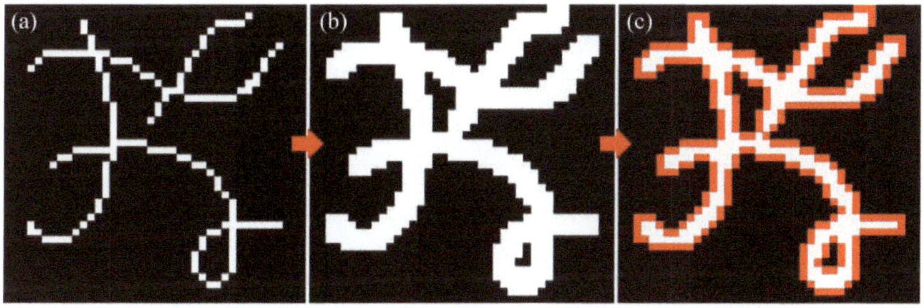

Fig. 5 Method of measurement of total length and number of DNAs [12].

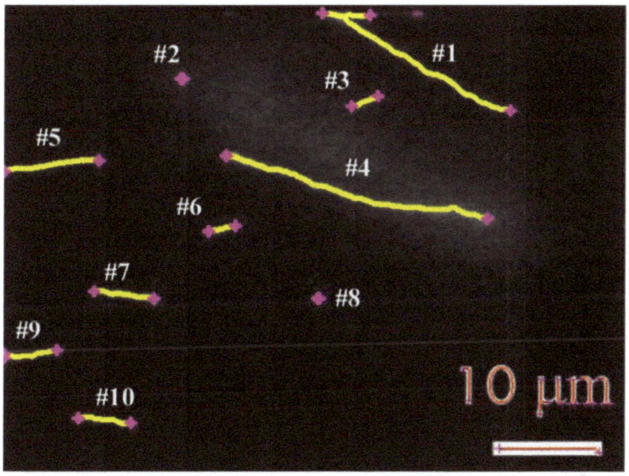

Fig. 6 Example of applying the measurement method of total length and number of DNA segments to an actual fluorescence microscope image [12].

Fig. 6 shows the image after the measurement method is applied to the image shown in Fig. 4A. In the image, the counters of the DNA segments are shown in yellow. The detected endpoints are shown by purple cross marks. Table 2 shows the comparison between DNA length measured by proposed image processing method, and which obtained by manual measurement by an operator. In manual measurement, the operator clicks many times on the image to approximate the DNA segments with polygonal lines to obtain the DNA length. The segment numbers in Table 2 correspond to the numbers shown in Fig. 6. The result shows that the measurement by proposed method provides almost the same length as the result of manual measurement although the value is slightly longer in most of the cases. The lengths of #2 and #8 are not measured in manual measurement because the lengths are too short. The difference is relatively large in the case of #1 because the part of the DNA segment protrudes from the image.

Table 2 Comparison between DNA length measured by proposed image processing method and which obtained by manual measurement [12].

Segment No.	Manual measurement [μm]	Measurement by image process [μm]	Difference [μm]
#1	19.87	23.70	3.82
#2	–	0.42	–
#3	3.32	2.96	−0.35
#4	25.63	26.32	0.69
#5	8.76	9.40	0.64
#6	2.75	2.79	0.04
#7	5.72	5.96	0.24
#8	–	0.33	–
#9	5.54	5.68	0.14
#10	6.83	5.31	−1.52

3.3 Image processing by deep learning

When measure the length and number of DNA segments by the proposed image processing method explained in Section 3.1, the operator must manually determine the tone curve in step-A3 and the control parameters shown in Table 1 to perform measurements on variety of images captured under different experimental conditions. In addition to that, operator must select DNA segments to measure because the images contain many images of impurities and curled DNA segments which should not be included for the calculation of average DNA length.

To achieve automatic measurement, deep learning models are developed for determining all controlled parameters and selecting measurable DNA segments. Fig. 7 shows the procedure of automatic measurement using deep learning models. The procedure uses five deep learning models: $\boldsymbol{M}_{\mathrm{param}}$, $\boldsymbol{M}_{\mathrm{conn}}$, $\boldsymbol{M}_{\mathrm{ext}}{}^{\mathrm{p}}$, $\boldsymbol{M}_{\mathrm{ext}}{}^{\mathrm{n}}$, and $\boldsymbol{M}_{\mathrm{noise}}$. The functions of these models are explained in Table 3. In step-B1, step-A1 to step-A6 are performed with predicted parameters by $\boldsymbol{M}_{\mathrm{param}}$. In step-B2, separate DNA segments, which should be considered as one DNA segment, are connected by $\boldsymbol{M}_{\mathrm{conn}}$ to improve the accuracy of DNA extraction in the next step. In step-B3a and step-B3b, the DNA extraction is performed by $\boldsymbol{M}_{\mathrm{ext}}{}^{\mathrm{p}}$ and $\boldsymbol{M}_{\mathrm{ext}}{}^{\mathrm{n}}$. To improve the accuracy, the two result images generated by two models $\boldsymbol{M}_{\mathrm{ext}}{}^{\mathrm{p}}$ and $\boldsymbol{M}_{\mathrm{ext}}{}^{\mathrm{n}}$ are superimposed in step-B4. To connect the DNA segments which are separated during the image processing, $\boldsymbol{M}_{\mathrm{conn}}$ is applied again after skeletonization and dilation in step-B5 to step-B7. In step-B8, the skeletonization is performed again. Then, the noise reduction $\boldsymbol{M}_{\mathrm{noise}}$ is applied.

Fig. 8 shows the architecture of $\boldsymbol{M}_{\mathrm{param}}$. Convolutional neural network (CNN) [20] is adopted for the model. $\boldsymbol{M}_{\mathrm{param}}$ received an equalized

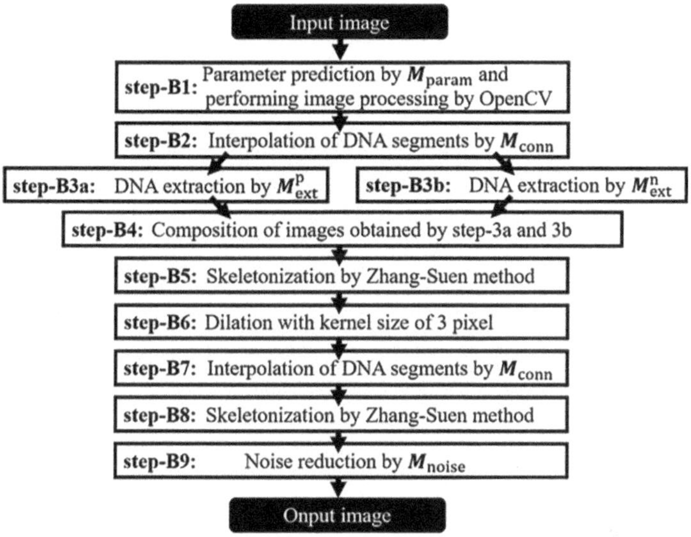

Fig. 7 Procedure of automatic image processing of DNA extraction by five deep learning models: $\boldsymbol{M}_{param}$, $\boldsymbol{M}_{conn}$, $\boldsymbol{M}_{ext}{}^{p}$, $\boldsymbol{M}_{ext}{}^{n}$, and $\boldsymbol{M}_{noise}$ [12].

Table 3 Explanation of five deep learning models used for the developed procedure shown in Fig. 7 [12].

Model	Functions
$\boldsymbol{M}_{param}$	Prediction of the control parameters and the tone curve used in step-A1 to step-A6
$\boldsymbol{M}_{conn}$	Connection of DNA segments that are separated in binarization of step-A4 due to the problem of local shading
$\boldsymbol{M}_{ext}{}^{p}$, $\boldsymbol{M}_{ext}{}^{n}$	Extraction of measurable DNA segments by removing impurities and curled DNA segments
$\boldsymbol{M}_{noise}$	Noise removal leaving the DNA segments

fluorescence microscope image (the image after step-A1) and output a 261-dimensional vector. Input images are resized to 512 × 512 pixels. The 5 of 261 dimensions correspond to the 5 parameters shown in Table 1, and remaining 256 dimensions correspond to the tone-curve in step-A3. All values of the output vector corresponding to integer parameters are scaled to 0–1.0 by dividing maximum of the range of the parameter. The values corresponding to bool values are set to 1.0 for Ture and 0.0 for False. 178 equalized fluorescence microscope images are prepared for the training and validation. Corresponding output vectors for the 178 images are prepared manually by an operator.

As explained earlier, image conversion is performed by $\boldsymbol{M}_{conn}$, $\boldsymbol{M}_{ext}{}^{p}$, $\boldsymbol{M}_{ext}{}^{n}$, and $\boldsymbol{M}_{noise}$ after the fluorescence microscope images are binarized

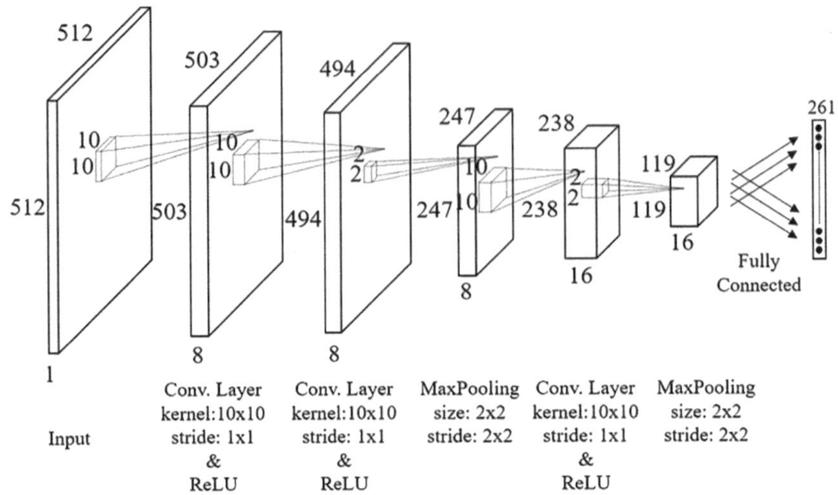

Fig. 8 Architecture of CNN adopted for $\boldsymbol{M}_{param}$ [12].

by step-A1 to step-A5 with predicted parameter provided by $\boldsymbol{M}_{param}$. $\boldsymbol{M}_{conn}$ interpolates the disconnected DNA segments. $\boldsymbol{M}_{ext}{}^{p}$ and $\boldsymbol{M}_{ext}{}^{n}$ extract DNA segments. $\boldsymbol{M}_{noise}$ reduces noise. Pix2pix [21], which is a deep learning model which generates a converted image from an input image, is adopted for the models of $\boldsymbol{M}_{conn}$, $\boldsymbol{M}_{ext}{}^{p}$, $\boldsymbol{M}_{ext}{}^{n}$, and $\boldsymbol{M}_{noise}$. Pix2pix is one of generative adversarial networks (GAN). The model consists of a generator $\boldsymbol{G}$ and a discriminator $\boldsymbol{D}$. In the case of pix2pix, $\boldsymbol{G}$ and $\boldsymbol{D}$ are trained by many of image pairs of $(\boldsymbol{x}, \boldsymbol{y}^{c})$, where $\boldsymbol{x}$ is an original input image and $\boldsymbol{y}^{c}$ is a corresponding converted image.

A software is developed to automatically generate enough sets of training data $(\boldsymbol{x}, \boldsymbol{y}^{c})$. The software generates imitation images of fluorescence microscope images which binarized by step-A1 to step-A5. In the software, the imitation images $\boldsymbol{x}$ and $\boldsymbol{y}^{c}$ are generated by three steps: drawing DNA segments; drawing impurities including curled DNA; drawing noise.

Due to the limited space available, only DNA extraction models: $\boldsymbol{M}_{ext}{}^{p}$ and $\boldsymbol{M}_{ext}{}^{n}$ are explained here. Please refer to Ref. [12] for $\boldsymbol{M}_{conn}$ and $\boldsymbol{M}_{noise}$. $\boldsymbol{M}_{ext}{}^{p}$ is trained to extract DNA segments from input images generated by the software. The input image $\boldsymbol{x}$ is generated by almost the same procedure of generation of $\boldsymbol{x}$ for the training of $\boldsymbol{M}_{conn}$ except for mixing cut DNA segments with uncut segments. By drawing pieces of cut DNA segments in $\boldsymbol{x}$ with uncut segments, $\boldsymbol{M}_{ext}{}^{p}$ is expected to learn the difference of the pieces of cut DNA segments from impurities which should be removed. The expected output images $\boldsymbol{y}^{c}$ is generated by drawing impurities and noise in gray.

In addition to $\boldsymbol{M}_{ext}^{p}$, $\boldsymbol{M}_{ext}^{n}$ is also trained to generate the images which only DNA segments removed. Using $\boldsymbol{M}_{ext}^{n}$, images of DNA segments are obtained by subtracting the output images $\boldsymbol{y} = \boldsymbol{G}(\boldsymbol{x})$ from the original input images $\boldsymbol{x}$. At the step-B4, two DNA extracted images obtained by $\boldsymbol{M}_{ext}^{p}$ and $\boldsymbol{M}_{ext}^{n}$ are superimposed. Both $\boldsymbol{M}_{ext}^{p}$ and $\boldsymbol{M}_{ext}^{n}$ have the probability to eliminate parts of DNA segments by misjudging as impurities. By superimposing two images obtained by different models, the probability of misjudgment can be reduced.

Fig. 9 shows the three examples of results of automatic image processing. Left images show the original input images obtained by actual fluorescence

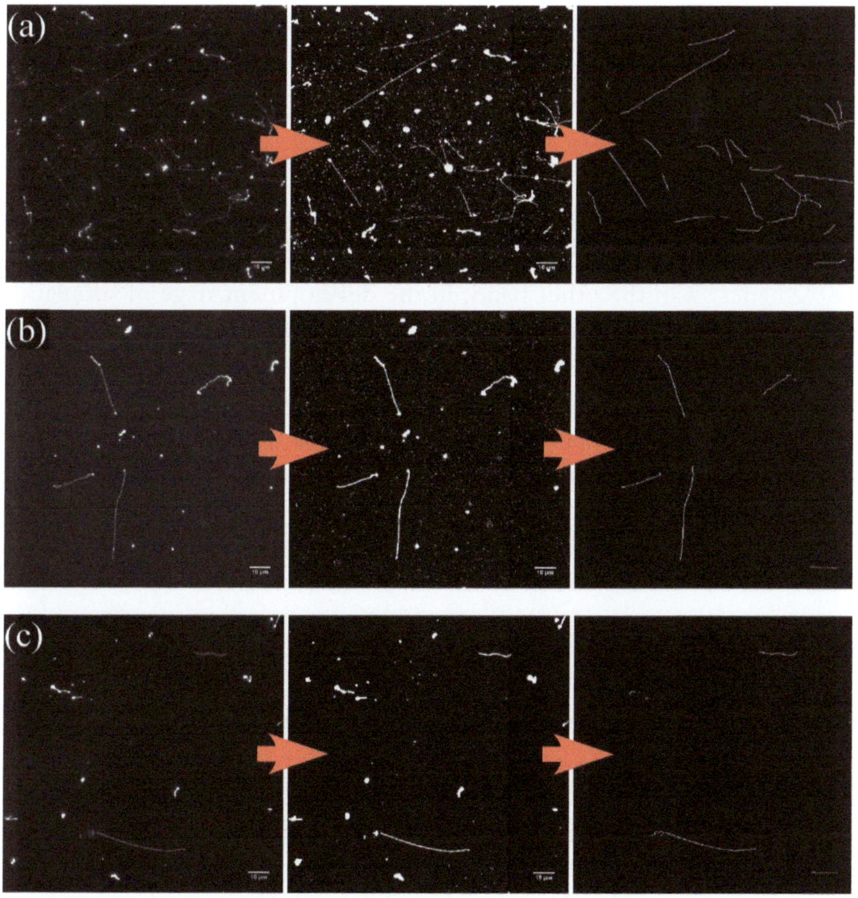

Fig. 9 Three examples of results of automatic image processing. Left images are the original input images obtained by actual fluorescence microscopes. Middle images are the image after step-B1. Right images are the final images after the processing (step-B9) [12].

microscope. Middle images are the image after step-B1 is applied. Right images show the final images after step-B9 is applied. Even for various input images with different contrast and brightness, DNA segments are almost correctly extracted after the parameters are automatically estimated, although some DNA segments that should be separated are connected, or some of faint parts of DNA segments are eliminated. By applying the measurement method explained in Section 3.2, total length of DNA segments can be measured.

4. In silico simulation of DNA damage by tritium β decay

4.1 Background

In recent years, computer simulation technology has made great progress, making it possible to calculate the motions and chemical reactions of large molecules such as DNA. Such advances in computational technology are expected to clarify details of DNA damage (e.g., presence or absence of cluster damage) and tritium transmutation effects (e.g., how much substitution can cause a problem?), which are difficult to elucidate only experimentally, and the effects of necrosis (e.g., how much substitution is problematic?). On the other hand, in the development stage of simulation technology, it will be necessary to compare the results of simulation with those of experiments on relatively simple systems to confirm the accuracy of the simulation.

From these perspectives, a computer simulation technique based on molecular dynamics has been developed to evaluate tritium-induced DNA damage, including transmutation effects [13–16]. In this section, computer simulation works on the structural changes of telomere DNA are reviewed.

4.2 Tritium decay effects on hydrogen bonds between telomere structures

All-atom simulations of a DNA molecule are difficult due to the performance of a computer. Therefore, the authors of Refs. [14–16] extracted the basic structure of telomeres, which are commonly found at the ends of DNA, and used them in their calculations. Telomeres are structures located at the ends of eukaryotic chromosomes that consist of repeats of specific base sequences (telomeric DNA, TTAGGGG in humans) and various proteins. There are four types of bases: adenine A, guanine G, cytosine C, and thymine T. Shortening of telomeres has been reported to cause atherosclerosis and cancer. The telomere DNA sequence used in Refs. [14–16] is d(TCTAGGGTTAGGGTTAG). The enzymes and proteins associated with telomeres were not considered in the computer simulation in Refs. [14–16], and solely the effect of tritium β-decay on telomere DNA was examined.

The tritium taken into the body is substituted by hydrogen from bases in telomeres and eventually β-decays to ^{3}He. Before the substitution, the hydrogen contributed to the structural formation of DNA by making covalent and hydrogen bonds with the surrounding atoms. However, because of the decay to ^{3}He, these bonds can no longer be maintained, and as a result, telomeric DNA structure collapse is feared.

Damaged DNA is repaired by enzymes through homologous recombination or non-homologous end joining [22]. In the former, homologous recombination, DNA sequences at sites of DNA sequence similarity (homologous sites) are used for precise repair. In the latter, nonhomologous end joining, the cleaved ends are directly linked to each other irrespective of DNA homology. Repair by nonhomologous end joining is less accurate because the reconnection may involve an increase or decrease in the number of nucleotides. In Ref. [14], the durability of telomere DNA alone against tritiated breakage was studied by quantitative MD simulations, limiting the effects to hydrogen bonds (Fig. 10).

Fig. 11 shows all the particles. To examine the effect of tritium transmutation on hydrogen bonding, we assume that the two hydrogen atoms in the NH_2 group of the guanine in Fig. 10 decay to two ^{3}He atoms. There are eight guanines in the telomere DNA used in Ref. [14]. The N guanine groups of DNA are decomposed. In other words, assuming that $2N$ hydrogen atoms decay to ^{3}He, the degree of β-decay is quantified by N.

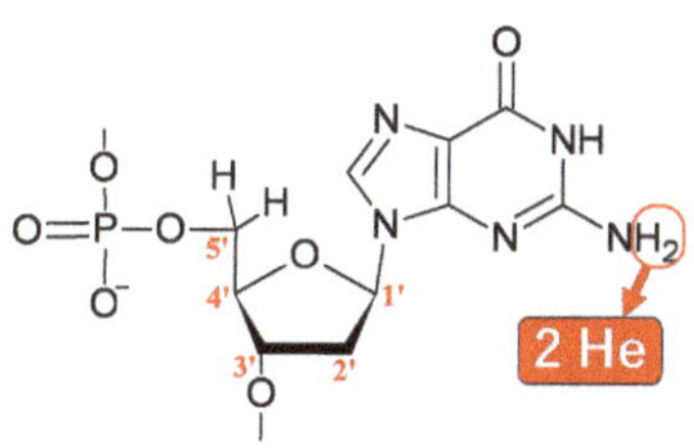

Fig. 10 Nucleotides that make up DNA. As a base, consider guanine (deoxyguanosine monophosphate). As noted in the red circles, part of hydrogen atoms in DNA are replaced by T atoms in the solvent those eventually β-decays to ^{3}He, and the effect was simulated using molecular dynamics [14].

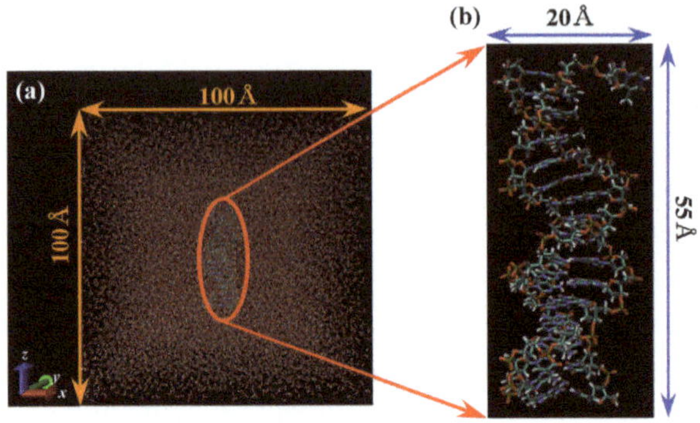

Fig. 11 (A) Telomere structure of human DNA. It exists in a solvent consisting of 71,000 water molecules, 121 Na^+ ions, and 89 Cl^- ions. Telomere DNA is composed of 1078 atoms. The simulation box is 100 Å × 100 Å × 100 Å. (B) Three-dimensional structure of telomere DNA in its initial state [14].

The time evolution of the structure with $2N$ hydrogen atoms is simulated by the NAMD code [23] using the CHARMM36 force field [24]. Temperature and pressure are controlled by the Nosé-Hoover Langevin piston method [25,26]. For details, see Ref. [14]. The time evolution of telomere structure for the case with the strongest β-decay ($N=8$) is shown in Fig. 12. The heat bath temperature is set to 310 K, assuming body temperature. As the time evolution progresses, the double strand is observed to unravel. The root mean square deviation (RMSD) is defined as an order parameter as follows, and its time evolution is illustrated in Fig. 13 for each of the cases $N=0$, 2, 4, 6, 8.

$$\mathrm{RMSD} = \sqrt{\frac{1}{n_0} \sum_{i=1}^{n_0} \left| \vec{r}_i(t) - \vec{r}_i^{\,\mathrm{PDB}} \right|^2}, \tag{5}$$

where i is the atoms constituting DNA excluding hydrogen, $1 \leqq i \leqq n_0$, ($n_0=691$). $\vec{r}_i(t)$ is the position vector of particle i at time t. $\vec{r}_i^{\,\mathrm{PDB}}$ is the initial position vector of particle i.

Fig. 13 shows that as N increases, i.e., as the degree of β-decay becomes stronger, the hydrogen bonds connecting the double chains weaken and the RMSD grows larger. In other words, the computer simulation could quantitatively evaluate the structure collapsing.

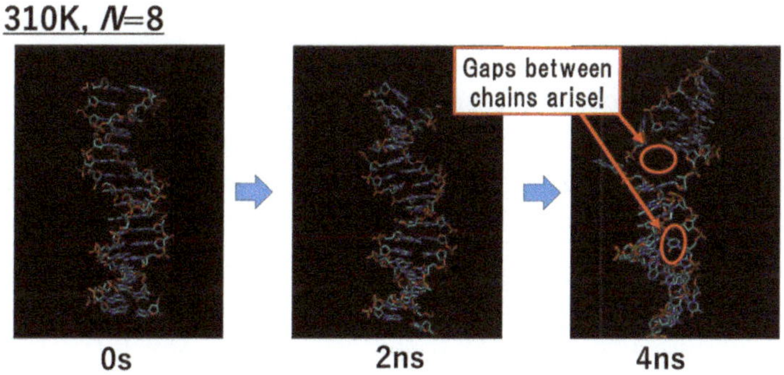

Fig. 12 Snapshot of the initial telomere structure ($N = 8$) as a starting point. The double-stranded structure "unravels" as time progresses. The temperature of the heat bath is set to 310 K, which is assumed to be body temperature [14].

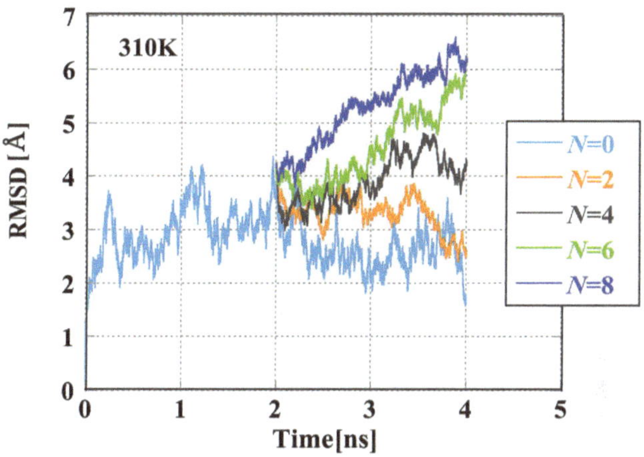

Fig. 13 Time evolution of the root mean square deviation (RMSD) of atomic positions for the telomere model to investigate the effect of hydrogen bonding. As the *N* increases, the RMSD becomes larger. It represents quantitatively the breakdown of the telomere structure.

5. Summary and future outlook

The rate of DSBs of DNA in tritiated water was quantitatively evaluated using a single molecule observation method. The effects of β-ray irradiation were not observable at tritium concentration of 1300 Bq/cm^3. The irradiation effects were clear at tritium concentrations of ~MBq/cm^3 and those were successfully separated from the contributions of other factors such as thermal motion and microorganisms. Comparison of the evaluated rates of radiation-induced DSBs with the values obtained by Monte-Carlo simulation using, for example, Geant4-DNA [27] and MD simulations are useful to check the accuracy of simulation and it is a future task. The image processing method for the automatic measurement of DNA length was successfully developed using OpenCV and deep learning. The quantitative evaluation using MD was performed for the effects of tritium transmutation on hydrogen bonds acting between the two main strands of DNA. A future task could be the investigation of the effect on covalent bonds within the main strand. The problem is the accuracy of the force field called ReaxFF, which can handle covalent bond breaking. Although there are several candidate force fields, a force field that describes a reasonable DNA has not been found at this point. It is necessary to search for the key force field for the evaluation of breakage of covalent bonds using MD simulations.

References

[1] United Nations Scientific Committee on the Effects of Atomic Radiation, United Nations Scientific Committee on the Effects of Atomic Radiation, Annex C, Biological effects of selected internal emitters-tritium, 2016. https://www.unscear.org/unscear/uploads/documents/publications/UNSCEAR_2016_Annex-C.pdf.

[2] S. Dingwall, C.E. Mills, N. Phan, K. Taylor, D.R. Boreham, Human health and the biological effects of tritium in drinking water: prudent policy through science—addressing the ODWAC new recommendation, Dose-Response 9 (2011) 6–31, https://doi.org/10.2203/dose-response.10-048.Boreham.

[3] S.B. Kim, N. Baglan, P.A. Davis, Current understanding of organically bound tritium (OBT) in the environment, J. Environ. Radioact. 126 (2013) 83–91, https://doi.org/10.1016/j.jenvrad.2013.07.011.

[4] L. Mullenders, M. Atkinson, H. Paretzke, L. Sabatier, S. Bouffler, Assessing Cancer risks of low-dose radiation, Nat. Rev. Cancer 9 (2009) 596–604, https://doi.org/10.1038/nrc2677.

[5] Y. Yoshikawa, M. Suzukia, N. Yamada, K. Yoshikawa, Double-strand break of giant DNA: protection by glucosyl-hesperidin as evidenced through direct observation on individual DNA molecules, FEBS Lett. 566 (2004) 39–42, https://doi.org/10.1016/j.febslet.2004.04.008.

[6] Y. Yoshikawa, T. Mori, N. Magome, K. Hibino, K. Yoshikawa, DNA compaction plays a key role in radioprotection against double-strand breaks as revealed by single-molecule observation, Chem. Phys. Lett. 456 (2008) 80–83, https://doi.org/10.1016/j.cplett.2008.03.009.

[7] Y. Ma, N. Ogawa, Y. Yoshikawa, T. Mori, T. Imanaka, Y. Watanabe, K. Yoshikawa, Protective effect of ascorbic acid against double-strand breaks in giant DNA: marked differences among the damage induced by photo-irradiation, gamma-rays and ultrasound, Chem. Phys. Lett. 638 (2015) 205–209, https://doi.org/10.1016/j.cplett.2015.08.054.

[8] M. Noda, Y. Ma, Y. Yoshikawa, T. Imanaka, T. Mori, M. Furuta, T. Tsuruyama, K. Yoshikawa, A single-molecule assessment of the protective effect of DMSO against DNA double-strand breaks induced by photo-and γ-ray irradiation, and freezing, Sci. Rep. 7 (2017) 8557, https://doi.org/10.1038/s41598-017-08894-y.

[9] T. Wada, A. Koike, S. Yamazaki, K. Ashizawa, F. Sun, Y. Hatano, H. Shimoyachi, T. Kenmotsu, T. Ikka, Y. Oya, Protective behavior of tea catechins against DNA double strand breaks produced by radiations with different linear energy transfer, Fusion Eng. Design 172 (2021) 112700, https://doi.org/10.1016/j.fusengdes.2021.112700.

[10] Y. Hatano, Y. Konaka, H. Shimoyachi, T. Kenmotsu, Y. Oya, H. Nakamura, Kinetics of double strand breaks of DNA in tritiated water evaluated using single molecule observation method, Fusion Eng. Design 146 (2019) 100–102, https://doi.org/10.1016/j.fusengdes.2018.11.045.

[11] Y. Hatano, H. Shimoyachi, T. Asano, T. Kenmotsu, T. Wada, Y. Oya, H. Nakamura, S. Fujiwara, Rate of double strand breaks of genome-sized DNA in tritiated water: Its dependence on tritium concentration and water temperature, J. Adv. Simulat. Sci. Eng. 9 (2022) 198–205, https://doi.org/10.15748/jasse.9.198.

[12] S. Saito, H. Nakamura, T. Kenmotsu, Y. Oya, Y. Hatano, Y. Tamura, S. Fujiwara, H. Ohtani, Image processing method for automatic measurement of number of DNA breaks, J. Adv. Simulat. Sci. Eng. 8 (2021) 173–193, https://doi.org/10.15748/jasse.8.173.

[13] S. Fujiwara, H. Nakamura, H. Li, H. Miyanishi, T. Mizuguchi, T. Yasunaga, T. Otsuka, Y. Hatano, S. Saito, Computational strategy for studying structural change of tritium-substituted macromolecules by a beta decay to helium-3, J. Adv. Simulat. Sci. Eng. 6 (2019) 94–99, https://doi.org/10.15748/jasse.6.94.

[14] H. Nakamura, H. Miyanishi, T. Yasunaga, S. Fujiwara, T. Mizuguchi, A. Nakata, T. Miyazaki, T. Otsuka, T. Kenmotsu, Y. Hatano, S. Saito, Molecular dynamics study on DNA damage by tritium disintegration, Jpn. J. Appl. Phys. 59 (2019) SAAE01, https://doi.org/10.7567/1347-4065/ab460d.

[15] H. Nakamura, S. Fujiwara, Y. Hatano, Molecular dynamics simulation on tritium-induced damage of biopolymers and experimental system for validation, J. Plasma Fusion Res. 97 (2021) 561–567 (in Japanese language) http://www.jspf.or.jp/Journal/PDF_JSPF/jspf2021_10/jspf2021_10-561.pdf.

[16] S. Fujiwara, Y. Hatano, H. Nakamura, Mechanism of DNA damage by tritium—fluorescence microscope observation of double-strand breaks and simulation, BUTSURI 77 (2022) 35–41 (in Japanese language).

[17] . https://imagej.nih/ij/.

[18] H. Shimoyachi, Quantitative Analysis and Mechanism of Double Strand Breaks of DNA by β-Rays from Tritium, Master thesis, University of Toyama, Japan, 2020.

[19] T.Y. Zhang, C.Y. Suen, A fast parallel algorithm for thinning digital patterns, Commun. ACM 27 (1984) 236–240, https://doi.org/10.1145/357994.358023.

[20] J. Gu, Z. Wang, J. Kuen, L. Ma, A. Shahroudy, B. Shuai, T. Liu, X. Wang, L. Wang, G. Wang, J. Cai, T. Chen, Recent advances in convolutional neural networks, arXiv (2017), https://doi.org/10.48550/arXiv.1512.07108.

[21] P. Isola, J.Y. Zhu, T. Zhou, A.A. Efros, Image-to-image translation with conditional adversarial networks, arXiv (2018), https://doi.org/10.48550/arXiv.1611.07004.

[22] A.A. Goodarzi, P.A. Jeggo, The repair and signaling responses to DNA double-strand breaks, Adv. Genet. 82 (2013) 1–45, https://doi.org/10.1016/B978-0-12-407676-1.00001-9.

[23] J.C. Phillips, R. Braun, W. Wang, J. Gumbart, E. Tajkhorshid, E. Villa, C. Chipot, R.D. Skeel, L. Kalé, K. Schulten, Scalable molecular dynamics with NAMD, Comput. Chem. 26 (2005) 17811802, https://doi.org/10.1002/jcc.20289.

[24] K. Hart, N. Foloppe, C.M. Baker, E.J. Denning, L. Nilsson, A.D. MacKerell Jr., Optimization of the CHARMM additive force field for DNA: improved treatment of the BI/BII conformational equilibrium, J. Chem. Theory Comput. 8 (2012) 348–362, https://doi.org/10.1021/ct200723y.

[25] G.J. Martyna, D.J. Tobias, M.L. Klein, Constant pressure molecular dynamics algorithms, J. Chem. Phys. 101 (1994) 4177–4189, https://doi.org/10.1063/1.467468.
[26] S.E. Feller, Y. Zhang, R.W. Pastor, B.R. Brooks, Constant pressure molecular dynamics simulation: the Langevin piston method, J. Chem. Phys. 103 (1995) 4613–4621, https://doi.org/10.1063/1.470648.
[27] http://geant4-dna.in2p3.fr/.

FSC
www.fsc.org
MIX
Papier | Fördert
gute Waldnutzung
FSC® C083411

Kolibri 360 GmbH
Ferdinand-Jühlke-Straße 7
99095 Erfurt, Deutschland
produktsicherheit@kolibri360.de